Demographic Diversity of the Science, Technology, Engineering, and Mathematics (STEM) Workforce in the U.S. Department of Defense

Analysis of Compensation and Employment Outcomes

JESSIE COE, MARIA C. LYTELL, CHRISTINA PANIS, WILLIAM SHELTON

Prepared for the Office of the Secretary of Defense
Approved for public release; distribution unlimited

RAND NATIONAL DEFENSE RESEARCH INSTITUTE

For more information on this publication, visit **www.rand.org/t/RRA1480-1**.

About RAND

The RAND Corporation is a research organization that develops solutions to public policy challenges to help make communities throughout the world safer and more secure, healthier and more prosperous. RAND is nonprofit, nonpartisan, and committed to the public interest. To learn more about RAND, visit www.rand.org.

Research Integrity

Our mission to help improve policy and decisionmaking through research and analysis is enabled through our core values of quality and objectivity and our unwavering commitment to the highest level of integrity and ethical behavior. To help ensure our research and analysis are rigorous, objective, and nonpartisan, we subject our research publications to a robust and exacting quality-assurance process; avoid both the appearance and reality of financial and other conflicts of interest through staff training, project screening, and a policy of mandatory disclosure; and pursue transparency in our research engagements through our commitment to the open publication of our research findings and recommendations, disclosure of the source of funding of published research, and policies to ensure intellectual independence. For more information, visit www.rand.org/about/principles.

RAND's publications do not necessarily reflect the opinions of its research clients and sponsors.

Published by the RAND Corporation, Santa Monica, Calif.
© 2023 RAND Corporation
RAND® is a registered trademark.

Library of Congress Cataloging-in-Publication Data is available for this publication.

ISBN: 978-1-9774-1093-1

Cover: hernan4429/Getty Images.

About This Report

The U.S. Department of Defense (DoD) continues to seek ways to better acquire and retain talent from the science, technical, engineering, and mathematics (STEM) disciplines to meet the technological challenges of existing and future missions. In the past, DoD (and the U.S. Congress) have focused on increasing STEM-worker compensation to attract and retain STEM talent. However, prior research reveals demographic-group differences in compensation within the STEM workforce, which could work against broader U.S. government goals to improve diversity, equity, inclusion, and accessibility in the federal civilian workforce.

This report documents findings and recommendations from a RAND National Security Research Division (NSRD) report to better understand demographic (gender, racial, and ethnic) group differences in compensation and career outcomes within DoD's civilian STEM workforce. This report builds on a fiscal year 2019 NSRD report that compared compensation between public- and private-sector STEM workforces (Edwards et al., 2021). As with the 2019 NSRD study, this report includes findings from quantitative analyses of DoD civilian personnel data. The authors first provide an overview of the composition of the DoD civilian STEM workforce. They then include findings on compensation differences between demographic groups while controlling for other observable characteristics, such as education, that might explain those group differences. Next, the authors describe findings from two exploratory excursions: first, into the compensation implications of the demographic composition of civilian pay plans, and second, into compensation differences while holding DoD component, geographic location, and STEM occupational category constant. The authors conclude with the key findings and recommendations for DoD to better understand and address potential demographic-related inequities in compensation within the DoD STEM workforce. The research reported here was completed in December 2022 and underwent security review with the sponsor and the Defense Office of Prepublication and Security Review before public release.

RAND National Security Research Division

This research was conducted within the Forces and Resources Policy Program of the RAND National Security Research Division (NSRD), which operates the RAND National Defense Research Institute (NDRI), a federally funded research and development center (FFRDC) sponsored by the Office of the Secretary of Defense, the Joint Staff, the Unified Combatant Commands, the Navy, the Marine Corps, the defense agencies, and the defense intelligence enterprise. This research was made possible by NDRI exploratory research funding that was provided through the FFRDC contract and approved by NDRI's primary sponsor.

For more information on the RAND Forces and Resources Policy Program, see www.rand.org/nsrd/frp or contact the director (contact information is provided on the webpage).

Acknowledgments

We extend our thanks to Molly McIntosh, director of the Forces and Resources Policy Program within NSRD, for her support throughout this project. We also thank Mallory Bulman from Gartner and Melanie Zaber from RAND for their thoughtful reviews of this report.

Summary

Issue

To develop and harness technological capabilities to meet its missions, the U.S. Department of Defense (DoD) continues to seek ways to better acquire and retain talent from the science, technical, engineering, and mathematics (STEM) disciplines. Congress and DoD policymakers point to higher compensation in the private sector as a key challenge. However, previous RAND Corporation research suggests that the average compensation difference between private- and federal-sector STEM workers is not that large after workforce characteristics are considered. This same research shows that there are demographic-group differences (gender, racial, and ethnic) in compensation for STEM workers. Given congressional and DoD interest in employing more STEM workers—and federal government interest in promoting diversity, equity, inclusion, and accessibility in the federal workforce more generally—demographic-group differences in the DoD STEM workforce warrant in-depth understanding. To that end, we build on previous RAND research to further explore demographic-group differences in compensation and, to a lesser degree, other employment-related outcomes (e.g., advancement) for the DoD civilian STEM workforce. Our aim is to better understand the demographic-group differences in these outcomes and to provide DoD with ways to ensure equitable outcomes across all workers in this critical workforce.

Approach

Using several years of data from the Defense Manpower Data Center (DMDC) that were provided to RAND, we quantify trends in demographic-group compensation differences among DoD civilians in the STEM workforce. We apply statistical models to account for other observable factors that could explain these differences, such as education level, DoD employer (component), geographic location (U.S. state), age, and pay plan. We then analyze compensation differences by pay plan and within select occupations in engineering sciences and information technology (IT) and computer science (CS). The available data limited our exploration of advancement and retention trends.

Key Findings

We found the following:

- White male workers have higher rates of unadjusted compensation than all demographic groups except for Asian men and women in the DoD STEM workforce.

- When workforce and organizational characteristics—such as education, age, tenure, and geographic location (U.S. state)—are included in our models, White men are compensated better than all other demographic groups. For example, White men make more than Asian men and women when location is taken into consideration.
- Within racial and ethnic groupings, women make less than their male counterparts.
- Controlling for workforce and organizational characteristics (i.e., education, age, tenure, and location) reduces compensation differences between White men and other groups except for Black women. If Black women were compensated for their characteristics in the same manner as White men, then they would receive an additional $7,500 annually on average.
- Location is particularly important for explaining the compensation differences among White men and Asian men and women.
- STEM worker compensation varies by pay plan: demonstration project (demo) pay plans have higher average compensation than General Schedule (GS) and similar pay plans. White men are overrepresented in demo pay plans; within GS and similar plans, they are overrepresented in the most-senior grades.
- After accounting for observable differences in worker and organizational characteristics, there is still a significant unexplained compensation difference among White men and all other demographic groups. This is especially prevalent when White men are compared with Black men and women. Although there are many potential influences, bias—either implicit or explicit—may exist. That is, our analyses cannot rule out the possibility that decisionmaking biases affect compensation and could factor into compensation differences.
- Our case study of Navy engineering and IT and CS occupations reveals that demographic-group differences in compensation still persist when geographic location (U.S. state) and pay grade are held constant. Using worker characteristics, we would expect Black women in Navy GS-12 IT management jobs in Virginia to make over $4,500 more than corresponding White men, whereas they actually earn $5,185 less on average. We would expect White women in Navy GS-12 engineering technician jobs in Virginia to make $849 less than White men; in reality, White women make $6,570 less than corresponding White men.

Recommendations

We offer three recommendations for DoD to better understand and address potential inequities in compensation in its civilian STEM workforce. For each recommendation, DoD should coordinate or communicate with the U.S. Office of Personnel Management, which is mandated by presidential executive order to review pay inequities in the federal civilian workforce. These recommendations are:

- *Establish additional guidance for DoD component use of alternative pay plans for DoD STEM workers.* Although an organization, such as the U.S. Department of the Navy, must apply to use a demo pay plan, it is not clear that it is required to benchmark compensation and other career outcomes of the STEM workers in that organization against all comparable STEM workers across the component. DoD can provide additional guidance for component-wide benchmarking to ensure that demo pay plan use is equitably applied.

- *Direct the services to assess STEM compensation and career outcomes by organization and location to better address unexplained demographic-group differences.* Given that civilian personnel decisions tend to happen at a local level, addressing any potential inequities for certain demographic groups likely requires a local-level review of policies and practices. We therefore suggest as a course of action that the Office of the Secretary of Defense (OSD) direct its service components to conduct assessments of its local organizations to understand how local policies, practices, and labor-market conditions affect demographic-group differences in compensation and other career outcomes. The service components would then report findings from the assessments to OSD to review.

- *Conduct additional analysis to understand the impact of entry-level compensation on demographic-group differences in compensation.* Because of resource and data limitations, we were not able to examine the impact of entry-level compensation, including differences in hiring bonuses and cost-of-living adjustments, on demographic-group differences in compensation. Examining where a DoD civilian STEM worker starts and is compensated initially could provide policymakers with valuable insights as to whether demographic-group differences in compensation start at the beginning of a career or emerge over time. Monitoring these entry-level compensation differences can also help policymakers determine whether trends are new or have been persistent over time.

Contents

Figures and Tables

Figures

Tables

Chapter 1. Introduction

National security agencies, particularly the U.S. Department of Defense (DoD), continue to struggle with acquiring and retaining talent from the science, technical, engineering, and mathematics (STEM) disciplines. Congress and federal agencies often assume that the key reason for this struggle is that private-sector employers provide much better compensation, making federal employment less attractive. Previous RAND Corporation research has found that compensation differences between the federal sector (including DoD) and private industry are, on average, not large when workforce characteristics such as education levels are taken into consideration (Edwards et al., 2021). However, the same RAND research shows that average compensation levels vary by demographic category and sector. For example, Asian workers have higher compensation in the private sector, White workers have similar compensation levels across federal and private sectors, and Black and Hispanic workers have lower compensation in the private sector than the federal sector. Regardless of racial and ethnic group, men have higher compensation than women.

Given congressional and DoD interest in employing more STEM workers—and federal government interest in promoting diversity, equity, inclusion, and accessibility (DEIA) in the federal workforce more generally—a more in-depth understanding of demographic-group differences in the DoD STEM workforce is warranted.[1] In particular, a better understanding of demographic trends in compensation and other employment outcomes (e.g., advancement) can help DoD tailor policies to ensure that its STEM workers have equitable compensation and career outcomes. To that end, we build on Edwards et al. (2021) by using available personnel data to further explore demographic-group differences in compensation and, to a lesser degree, other employment-related outcomes (e.g., advancement) for the DoD civilian STEM workforce. Our aim is to better understand disparities in how different demographic groups are compensated

---

[1] For example, in July 2021, President Joseph R. Biden, Jr., issued an executive order to revitalize efforts to promote DEIA in the federal workforce (Biden, 2021). This order defines DEIA terms accordingly:

- *Diversity* is "the practice of including the many communities, identities, races, ethnicities, backgrounds, abilities, cultures, and beliefs of the American people, including underserved communities."
- *Equity* is "the consistent and systematic fair, just, and impartial treatment of all individuals, including individuals who belong to underserved communities that have been denied such treatment."
- *Inclusion* is "the recognition, appreciation, and use of the talents and skills of employees of all backgrounds."
- *Accessibility* is "the design, construction, development, and maintenance of facilities, information and communication technology, programs, and services so that all people, including people with disabilities, can fully and independently use them."

and to provide DoD with recommendations on how to ensure equitable outcomes for all workers in this critical workforce.[2]

Our Approach

We take an econometric approach to explore demographic-group differences in compensation and other career outcomes. For our analyses, we leveraged several years of administrative personnel data on DoD civilians provided to RAND by the Defense Manpower Data Center (DMDC).

First, we describe the gender and racial and ethnic distribution of the DoD civilian STEM population by workforce and organizational factors (e.g., education levels, DoD employer). Specifically, we used 11 years (fiscal years [FYs] 2010–2020) of DMDC data to quantify the gender and racial and ethnic distribution of DoD civilian STEM workers by education level, STEM occupational category, DoD employer (DoD overall and U.S. Army, Navy, and Air Force), and pay plan.[3] The findings from this analysis are presented in the next section of this chapter.

Second, we quantify demographic-group differences in compensation, advancement, and retention. We first quantified average levels of demographic-group differences in outcomes over time. Depending on the outcome, we used different years of data. We provide the rationales for these data choices later in the report. For compensation specifically, we applied regression models and an econometric decomposition analysis to understand what is driving observed differences in compensation by demographic group. Consider a notional example of two DoD civilian STEM workers, Bill and Tina. Bill is a White man with a college degree working in computer science (CS). He is 30 years old and has been at his job for five years. Tina is a Hispanic woman with a Ph.D. working in life sciences. She is 30 years old and has been at her job for three years. In this example, we observe that Bill has higher compensation than Tina. Although this scenario could exemplify racial or ethnic bias, it also can be explained by one worker's longer tenure and work in a STEM area that tends to be better compensated than other STEM areas. We employed regression models to investigate the extent to which differences in observed characteristics explain differences in compensation using gender, race, and ethnicity.

---

[2] Edwards et al. (2021) provide additional context to the differences in compensation between the private and public sectors, including those related to demographic characteristics, such as age, gender, race, and ethnicity. Our focus is solely on differences within the DoD civilian STEM workforce, but we acknowledge that trends in the STEM workforce outside DoD will affect DoD's ability to attract and retain STEM talent.

[3] We used the five STEM categories defined and used by Edwards et al. (2021): engineering science, information technology (IT) and computer science (CS), life science, physical science, and social science. See Edwards et al. (2021) for a detailed description of the occupational categories that fall within each STEM category.

A *pay plan* refers to "a particular table or array of pay rates prescribed by law or other authoritative source that establishes the basic pay rates for certain employees" (U.S. Office of Personnel Management [OPM], undated-a). For more details on the pay plans in the DoD workforce, see the section in this chapter entitled "Pay Plan."

Decomposition analysis allows us to further investigate whether demographic groups are paid differently for the same characteristics and to quantify fully unobservable compensation differences between them.[4] In the case of Bill and Tina, regression analysis would account for the differences in education, STEM category, and tenure, all of which affect compensation. It also can indicate whether demographic identity is correlated with any remaining difference in compensation. The decomposition analysis might reveal that, although Bill and Tina are the same age, 30-year-old Hispanic women are paid relatively better than 30-year-old White men. This is a difference in rate of pay based on a characteristic, also called a *return to a characteristic*, as opposed to a difference in the characteristic itself (for which the regression analysis can account). The decomposition analysis also quantifies the pay gap that is not explained by either the differences in characteristics or the returns to characteristics—what we call the *unexplained difference*. Although we cannot account these differences, we can quantify them across demographic groups. These can be illustrative and informative for future research.

Our initial research plan proposed a model-based analysis of advancement and retention outcomes. However, the average rates of retention were very high and similar among demographic groups, and the structural differences between civilian pay plans (i.e., pay grades or bands) made it difficult to assess advancement. Instead of looking across pay plans, we looked within a few select pay plans. This resulted in small numbers for most demographic groups, which hindered our ability to make meaningful model-based analyses. Instead, we pivoted to assessing pay plan assignment by demographic group and demographic-group representation in the top pay grade within a given pay plan. For advancement and retention, we offer descriptive details in Appendix A.

Third, we explore demographic-group composition of pay plans and top pay grades. Because of the importance of pay plans in setting compensation levels, we look further at demographic-group differences in compensation within six pay plans that are commonly used in the DoD civilian STEM workforce. We employed regression analysis to explore pay plan assignment by demographic group and representation in the top pay grade within pay plan.

Finally, using available data and sufficient samples sizes of the demographic groups, we take a deeper dive into trends within two large occupational categories in the DoD STEM workforce: engineering sciences and IT and CS. We focus on how compensation differences relate to two STEM occupational categories, holding agency, location (U.S. state), and STEM category constant, which results in holding job characteristics constant. We take this case-study approach because our models revealed differences in how White men are compensated compared with women and other races and ethnicities that were not fully explained by age, education level, tenure, location, agency, or STEM category. Our deeper dive focuses on the Navy because it

---

[4] *Fully unobservable* here means differences in compensation that cannot be attributed to either differences in the characteristics included in the model (e.g., education) or different pay rates for those characteristics (e.g., male workers with a Ph.D. paid higher than female workers with a Ph.D.).

employs enough personnel from each demographic group in a single location and occupational category.

Demographic Composition of the Department of Defense STEM Workforce

In Figures 1.1 and 1.2, we provide a high-level description of the demographic (gender and racial and ethnic) composition of the DoD STEM workforce. We consider eight demographic groups defined by the cross of binary gender and four racial and ethnic categories. The race and ethnicity categories that we consider are Asian, Black, Hispanic, and White.[5] Figure 1.1 shows the demographic composition of the DoD STEM workforce in FY 2020. Figure 1.2 shows how this composition has changed over time.

Consistent with Edwards et al. (2021), we find that the DoD STEM workforce is majority White male (61 percent in FY 2020).[6] The demographic composition has remained stable since FY 2010, with a slight decrease in the percentage of White men.

We provide descriptions of the workforce by educational attainment, STEM category, DoD employer, and pay plan. These descriptive details help provide a baseline for our more complex analyses later in the report.

---

[5] Asian, Black, and White are used as shorthand for non-Hispanic Asian, non-Hispanic Black, and non-Hispanic White, respectively. *Non-Hispanic Asian* includes Asian as well as Pacific Islander and Hawaiian Asian. Workers with race or ethnicity listed as "other" or missing are excluded from the analysis. The resulting four racial and ethnic categories that we consider encompass 97 percent of the DoD STEM workforce between FYs 2010 and 2020.

[6] This general trend is also consistent with the U.S. STEM workforce. For example, in 2019, the U.S. STEM workforce was 66 percent male and 65 percent White (Okrent and Burke, 2021).

Figure 1.1. Demographic Composition of the Department of Defense STEM Workforce

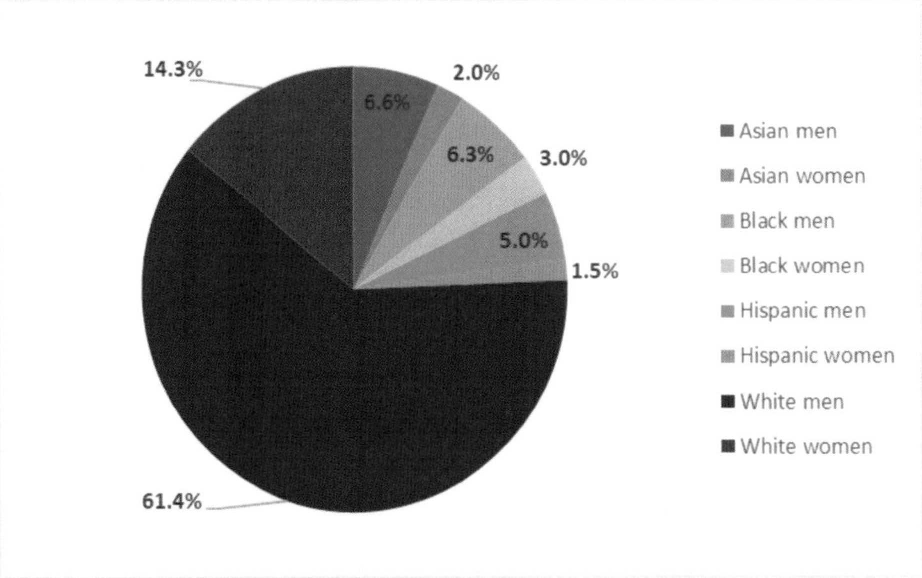

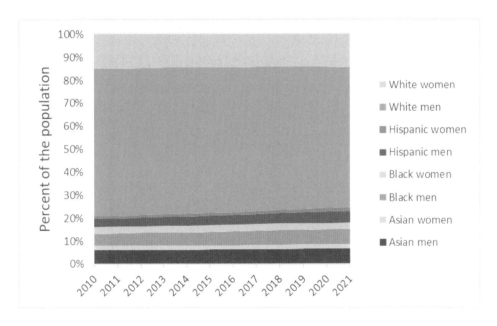

SOURCE: Authors' analysis of DMDC data provided to RAND on DoD civilian STEM workers (FY 2020).
NOTE: Percentages do not sum to 100 because of rounding.

Figure 1.2. Changes in the Demographic Composition of the STEM Workforce in the Department of Defense Over Time

SOURCE: Authors' analysis of DMDC data provided to RAND on DoD civilian STEM workers (FYs 2010–2020).

Education

Educational attainment also varies by race and ethnicity among DoD STEM workers. As shown in Figure 1.3, Asian workers have the highest educational attainment; more than

80 percent have a post-secondary degree or certification (all bar segments except those marked in green). White, Black, and Hispanic have similar trends: There are somewhat higher proportions of Black workers with master's degrees (bar segments marked in orange).

Figure 1.3. Educational Attainment by Demographic Group for the Department of Defense STEM Workforce

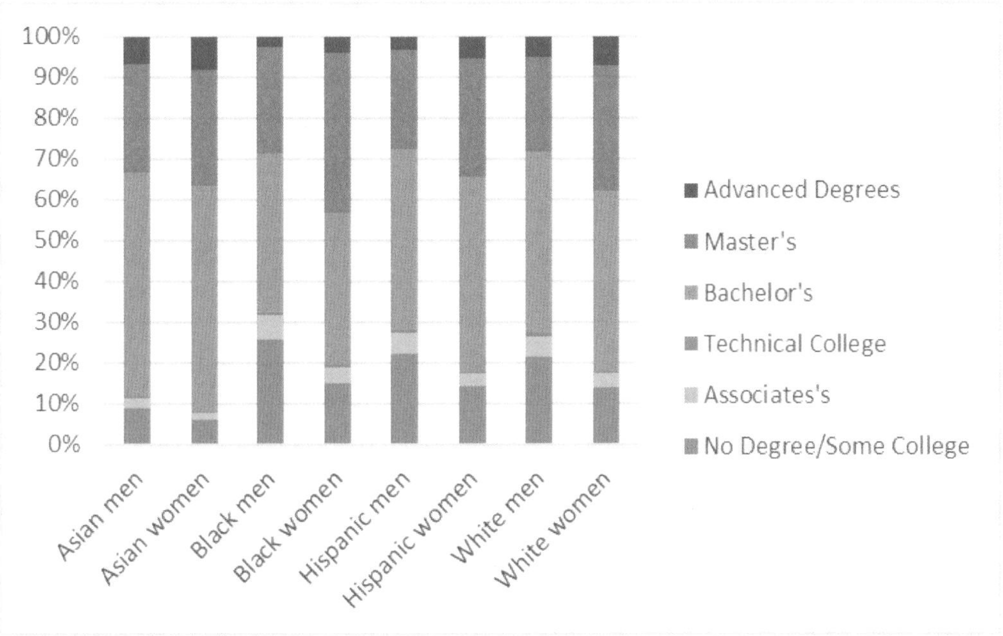

SOURCE: Authors' analysis of DMDC data on DoD civilian STEM workers (FY 2020) provided to RAND.

STEM Category

We follow Edwards et al. (2021) to classify STEM workers into five distinct STEM categories: social science, engineering science, IT and CS, life science, and physical science. The report showed that average compensation varies by these five STEM categories within the federal government and private sector. Also, the STEM categories vary in their demographic distribution, and this variance can manifest as demographic-group differences in compensation. We show the demographic composition of the DoD STEM workforce by STEM category in Figure 1.4. For example, nearly 70 percent of Asian men are in engineering sciences, whereas over 50 percent of Black men and over 50 percent of Black women are in IT or CS jobs.

Figure 1.4. STEM Category by Demographic Group for the Department of Defense STEM Workforce

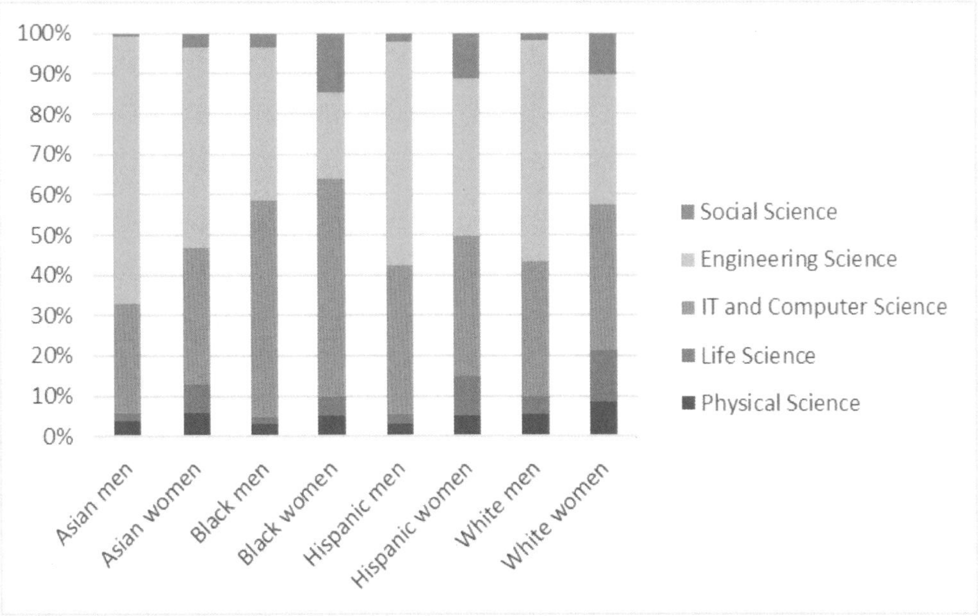

SOURCE: Authors' analysis of DMDC data on DoD civilian STEM workers (FY 2020) provided to RAND.

Employer

Figure 1.5 presents the demographic distribution of DoD STEM workers across the four main departments in the DoD workforce: U.S. Department of the Navy, U.S. Department of Defense (headquarters and fourth estate), U.S. Department of the Army, and U.S. Department of the Air Force.[7] The Navy and Army employ the most DoD STEM workers, although department representation varies by demographic category. Over half of Asian men and over half of Asian women work for the Navy, with White men following at about 45 percent. Other than Asian men and Asian women, at least 30 percent of STEM workers in each of the other demographic groups work for the Army.

---

[7] The "fourth estate" includes any DoD organizations that do not fall within a military branch or a combatant command (DoD, 2019). There are over 30 such organizations throughout DoD.

Figure 1.5. Employer by Demographic Group for the Department of Defense STEM Workforce

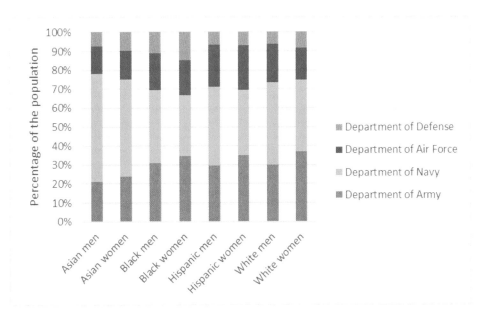

SOURCE: Authors' analysis of DMDC data on DoD civilian STEM workers (FY 2020) provided to RAND.

Pay Plan

The U.S. Office of Personnel Management (OPM) defines a *U.S. federal civilian workforce pay plan* as a "particular table or array of pay rates prescribed by law or other authoritative source that establishes the pay rate for certain employees" (OPM, undated-a). Each pay plan includes different job levels or grades at which compensation is set and often includes other personnel management features, such as special authorities for hiring. Thus, a pay plan not only sets the compensation structure for a particular workforce but has implications for how personnel in that plan are hired and managed.

The most used and perhaps best-known federal civilian pay plan is the General Schedule (GS) plan. Its regulations and policy are overseen by OPM. The GS plan covers a wide variety of civilian occupations and has 15 grades (GS-01 to GS-15). The defense intelligence community uses a pay plan with a similar structure called GG (General Government). Because of their similar grade structure, our analysis combines GS and GG.

The DoD STEM workforce is also compensated through other pay plans besides GS and GG. The most prominent among these are known as *demonstration project* (or *demo*) plans, which were established to flexibly manage and compensate workers with critical skills, such as STEM training. There are two main categories of demonstration project plans in DoD: Civilian Acquisition Workforce Personnel Demonstration (AcqDemo) and Scientific and Technical Reinvention Laboratories Demonstration (Lab Demo). Demo plans use pay bands instead of pay grades; pay bands are wider than grades and allow managers to pay personnel for performance (e.g., assign bonuses) without having to promote them to the next band. Demo plans also have other authorities, such as direct hire, which waives certain federal hiring requirements and speeds up the hiring process. These types of authorities, as well as other demo-plan features, such as performance-based bonuses, may result in different compensation levels for demo pay plans compared with GS and GG plans. Although a full discussion of these pay plans is beyond the scope of this report, we refer interested readers to Groeber et al. (2020).

The top pay plans in the DoD STEM workforce are shown in Table 1.1. Although the majority of DoD STEM workers are in GS or GG pay plans (64.7 percent), a sizable portion are in demo plans, such as NH (an AcqDemo plan for business management and professionals) and ND (a Lab Demo plan).

Table 1.1. Top Pay Plans for the Department of Defense STEM Workforce

Pay Plan	Number of DoD STEM Workers	Percentage of DoD STEM Workers
DB: Demonstration engineers and scientists (Army)	6,852	4.7
DP: Demonstration professional (Navy)	5,554	3.8
DR: Demonstration Air Force scientist and engineer	2,708	1.9
GG and GS	92,410	63.7
ND: Demonstration scientific and engineering	14,499	10.0
NH: Business management and technical management professional	11,223	7.7
NT: Demonstration administrative and technical	3,074	2.1
All others	8,975	6.1

SOURCE: Authors' analysis of DMDC data on DoD civilian STEM workers (FY 2020) provided to RAND.

Figure 1.6 shows the split between the GG and GS pay plans and all other pay plans by demographic group. Asian men and White men are much more likely to have alternative pay plans to GG and GS than other demographic groups; both Black men and women, as well as Hispanic women, are overwhelmingly on GG or GS pay plans.

Figure 1.6. Civilian Pay Plan Distribution by Demographic Group for the Department of Defense STEM Workforce

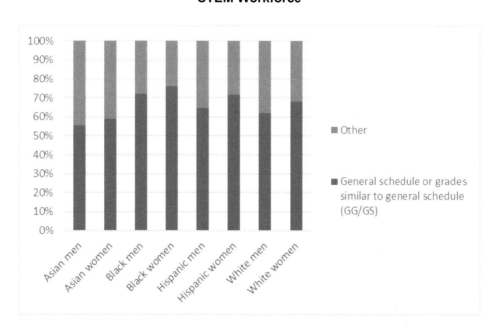

SOURCE: Authors' analysis of DMDC data on DoD civilian STEM workers (FY 2020) provided to RAND.

These differences in composition have implications for compensation. Higher levels of education are correlated with higher levels of compensation, and different STEM categories have different levels of average compensation. As we will show in Chapter 3, different pay plans have different average levels of compensation: GG and GS plans generally have lower levels of compensation than other pay plans. To analyze compensation disparity across demographic groups, we must account for these different characteristics together. We turn to that task in the next chapter.

Limitations

Our analysis is largely quantitative in nature. Although we cite a few previous studies to supplement our conclusions, we do not provide a comprehensive review of literature. We have built our research on the Edwards et al. (2021) study, which includes a detailed literature review of these topics; we refer readers to that report.

While our analysis is quantitative, it should be viewed more as descriptive than causal. This is partly because breaking down results by demographic group yielded population sizes that were too small for quantitative analysis and partly because of a lack of all relevant variables to explore the relevant mechanisms. Where feasible, we apply statistical models to control for observable characteristics (other than gender, race, and ethnicity) but acknowledge that other characteristics not included in our models (such as parental status) could help explain demographic-group differences in compensation.

We also limited our scope to examining outcomes for those already in the DoD STEM workforce. Specifically, we do not examine data on applicants for DoD STEM positions nor do we follow those who leave the workforce. Previous RAND research that analyzes applicant data for DoD civilians has noted the data limitations; Matthews et al. (2017) recommended that DoD review applicant data to ensure their accuracy. We also chose to examine compensation in-depth, but we did not examine entry pay grades (or bands). However, previous RAND research suggests that pay grade at entry can have important implications for women's advancement compared with men (Keller et al., 2020).

Structure of This Report

The rest of the report presents findings aimed at explaining compensation differences by demographic groups in the DoD civilian STEM workforce. In Chapter 2, we analyze compensation across the eight demographic groups for the entire DoD STEM workforce. In Chapter 3, we examine select pay plans commonly used in the DoD STEM workforce and delve farther into pay plan differences within pay grades. In Chapter 4, we explore occupational differences that might explain differences in compensation across demographic groups but for a fixed location, agency, and STEM category. Chapter 5 presents our key findings and recommendations to better understand and address potential inequities in compensation in DoD's civilian STEM workforce.

We also include three appendixes. Appendix A presents average advancement rates and retention rates by demographic groups; Appendix B includes additional results from our decomposition analysis of compensation differences in Chapter 2; and Appendix C shows results on pay plans and occupations to supplement findings in Chapters 3 and 4, respectively.

Chapter 2. Compensation

Our main analysis focuses on compensation differences by gender as well as race and ethnicity within the DoD civilian STEM workforce. As described in the introduction, we classify workers using the cross between binary gender (male, female) and race and ethnicity (Asian, Black, Hispanic, and White). Our measure of compensation is gross pay, which includes basic pay as well as any premium or other pay. Throughout this section we will refer to *compensation* as compensation, wages, and earnings interchangeably.

Figure 2.1 shows compensation trends for our eight demographic groups. The left panel uses annual dollars (nominal), and the right panel is inflation-adjusted (real) by the consumer price index to 2021 dollars. Asian male workers are the highest earners. Asian female workers and White male workers have similar levels of compensation. In contrast, Edwards et al. (2021), using Current Population Survey (CPS) data from 2018, found that White male workers were the highest paid demographic group and Asian female workers were actually the lowest paid demographic group.[8] The administrative data reveal a different ordering of pay by demographic group than what was found in public data and have important implications for understanding gender as well as racial and ethnic differences in compensation. DMDC data on the DoD STEM workforce shows, for example, that Black and Hispanic women are the two lowest-paid demographic groups: Hispanic women have lost ground starting in 2015 and were the lowest-paid demographic group in 2020.

To obtain the most-relevant results to current managerial practices, the rest of this chapter limits the time frame to FYs 2018–2020. Furthermore, the stability of compensation trends over those three years allows us to show how compensation differences are driven by group as opposed to relative differences over time.[9]

---

[8] The CPS is jointly administered by the Census Bureau and Bureau of Labor Statistics to collect "labor force statistics for the population of the United States" (U.S. Census Bureau, 2022). The survey is administered monthly to a "probability selected sample of about 60,000 occupied households" (U.S. Census Bureau, 2021).

[9] Other factors that narrowed our time frame were the federal governmentwide policies that resulted in federal civilian pay freezes, hiring freezes, unpaid days off, and mandated increases to contribution rates for civilian retirement plans from 2011 to 2013. A RAND analysis of these policies showed that long-term pay freezes would be expected to decrease retention of DoD civilians with at least bachelor's degrees (Asch, Mattock, and Hosek, 2014). Another federal civilian hiring freeze occurred and ended in 2017 (Naylor, 2017). We acknowledge that the federal government shutdown from December 2018 through January 2019 also may have had effects on federal civilian employment patterns. As we note here, the compensation trends appear fairly stable in our data during the 2018–2021 time frame. We cut off at FY 2020 because that is the last year for which we have complete data.

Figure 2.1. Compensation of STEM Workers by Demographic Group Over Time

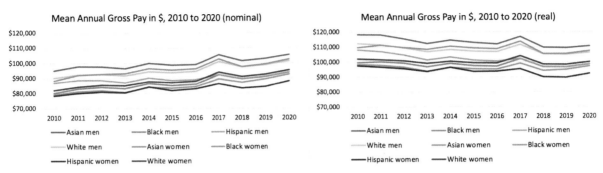

SOURCE: Authors' analysis of DMDC data on DoD civilian STEM workers (FYs 2010–2020) provided to RAND.
NOTE: Real numbers in the right-hand panel refer to inflation-adjusted estimates using consumer price index for 2021 dollars. Nominal amounts in the left-hand panel give dollar amounts in current year.

Regression Models Account for Observable Characteristics That Explain Demographic-Group Differences in Compensation

Comparing compensation across demographic groups suggests some discrepancies using gender as well as race and ethnicity, but other characteristics that influence compensation vary by demographic group (as noted in Chapter 1). For example, Asian workers are more likely to hold advanced degrees and more likely to be in the engineering category, both of which are associated with higher compensation levels.

We use regression analysis to control for compositional differences across demographic groups. Regression analysis estimates the association between an outcome variable (in this case, compensation) and a variable of interest (in this case, demographic group), holding other variables constant. Regression analysis gives an estimate of the average difference in compensation across demographic groups assuming that the composition of the groups was the same. The raw differences shown in Figure 2.1 encompass both disparities in compensation because of group composition (worker's education or agency) and differences associated with gender and race and ethnicity. Regression analysis separates out the differences using composition and estimates the differences based solely on demographic groups. It is important to note that any remaining differences based on demographic group are not necessarily evidence of discrimination. Regression analysis allows us to separate out the effect of observed differences across groups, but there are many unobserved factors that affect compensation that may vary by demographic group—such as worker preferences and job quality.

Table 2.1 lists the observed characteristics included in our regression models, including worker age, education level, location, and tenure; as well as job STEM category, agency, and pay plan.[10] Location is not included in the table because of space. "Observations" are the total number of observed characteristics between FYs 2018 and 2020. "Percentage" lists the percentage represented by that characteristic. For example, 23.6 percent of the observations are workers between ages of 18 and 35 years old. Both age and length of service are split into roughly equal groups. "Average Gross Pay" is the average compensation for that characteristic. For example, the average compensation level for a STEM worker with an advanced degree is $123,942 a year. Higher levels of education, longer career times, and higher ages are all associated with higher levels of compensation.

---

[10] We do not list pay plan details in Table 2.1 because of space considerations. This chapter uses all pay plans available in the data, and we allow for different average levels of compensation for each pay plan. In the following chapters, we limit our analysis to select pay plans.

Table 2.1. Average Compensation of Department of Defense STEM Workers

Category	Level	Observation	Percentage (%)	Average Gross Pay ($)
All			100.0	97,336
Education	No degree/some college	82,117	19.3	84,984
	Associate's degree	19,537	4.6	85,575
	Technical college	2,789	0.7	88,946
	Bachelor's degree	190,524	44.7	94,618
	Master's degree	108,580	25.5	108,287
	Advanced degree	22,416	5.3	123,942
Age	18 to 35 years	100,281	23.6	71,355
	35 to 45 years	102,794	24.2	96,074
	45 to 55 years	105,624	24.9	105,130
	55+ years	116,327	27.4	113,628
Tenure	0 to 2 years	43,879	10.3	50,057
	2 to 5 years	53,462	12.6	79,928
	5 to 10 years	79,667	18.7	91,542
	10 to 15 years	82,677	19.4	101,539
	15 to 20 years	60,412	14.2	109,292
	20+ years	105,865	24.9	119,979
STEM category	Physical science	23,653	5.6	101,696
	Life science	23,087	5.4	89,531
	IT and CS	147,991	34.7	94,716
	Engineering science	215,811	50.7	100,244
	Social science	15,421	3.6	86,793
Agency	Air Force	81,716	19.2	87,335
	Army	130,294	30.6	96,270
	DoD, fourth estate	31,209	7.3	106,970
	Navy	182,744	42.9	100,923

SOURCE: Authors' analysis of DMDC data on DoD civilian STEM workers (FYs 2018–2020) provided to RAND.
NOTE: Annual gross pay is in 2021 dollars.

Using White men as the reference, Figure 2.2 presents both the average raw differences in compensation by demographic group, and the average differences in compensation by demographic group adjusted for other observed characteristics. The differences are presented in both real dollar amounts and percentages. A raw difference of –7.6 percent for White women thus indicates that White women made an average of 7.6 percent less than White men between FYs 2018 and 2020. The regression results are the adjusted differences. The full regression estimate includes age, length of service, education level, agency, state, STEM category, and pay plan as controls and can be interpreted as the average difference in compensation between a given group and White men, holding the controls constant. The *average* aspect refers to both the average across members within demographic groups and across the different levels of the control variables. For example, there may be differences in compensation between White women and White men with bachelor's degrees who are both 35 years old and work in IT and CS for the Navy in California. There also may be differences in compensation between White women and White men with advanced degrees who are 40 years old and work in engineering for the Air Force in Maryland. Each of those examples holds the control variables fixed, but at different levels. The regression coefficient on White women can be thought of as the average across all the possible levels of control variables which affect the difference in compensation between White women and White men.

Figure 2.2. Raw and Adjusted Compensation Differences by Demographic Group

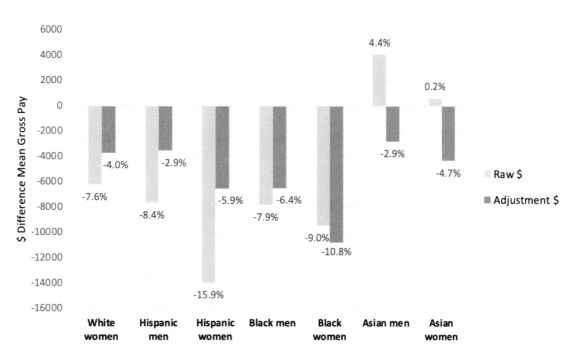

SOURCE: Authors' analysis of DMDC data on DoD civilian STEM workers (FYs 2018–2020) provided to RAND. NOTE: Raw results are from a regression including only year controls. Adjusted results include age, length of service, education level, state, agency, STEM category, and pay plan as controls. Outcome is annual compensation in 2021 dollars.

Within every demographic group, women earn less than men both in terms of raw compensation and when fully adjusted for observed differences. Fully adjusted for observed differences, White women earn 4 percent less than White men, Hispanic women earn 3 percent less than Hispanic men (–5.9 percent versus –2.9 percent), Black women earn 4.4 percent less than Black men (–10.8 percent versus –6.4 percent), and Asian women earn 1.8 percent less than

Asian men (–4.7 percent versus –2.9 percent). All these differences are statistically significantly different than zero at the one percent level.[11]

The average difference in compensation between White women and their male counterparts drops from 7.6 percent lower to 4 percent lower when fully adjusted. This indicates that variations in group composition, like STEM category (indicated in Figure 1.4), explain some of the difference in compensation. However, there remains a 4 percent pay gap that is not explained by the observed characteristics. On average, Asian men and women receive higher compensation than White men in STEM, but once observed group differences are accounted for, Asian men and women actually earn less. Table 2.2 provides the full regression results. Column (1) shows the raw results that only account for annual changes in overall compensation. Column (7) displays the fully adjusted results. The columns in between successively add control variables as indicated by the "x" entries in the bottom panel. When location is included in the controls, a large shift happens to the results for Asian men and women. This indicates that differences in location explain why Asian workers have higher compensation than White male workers. When comparing workers in the same area, Asian workers earn less than their White male counterparts by thousands of dollars on average.

For both Hispanic male and female workers, differences in composition compared with White men explain much of the raw differences in compensation. For Hispanic men, observed differences explain 5.5 percentage points of the –8.4 percent difference in raw compensation compared with White men, which corresponds to over $4,100 of the $7,649.[12] The remaining $3,513 difference between Hispanic men and White men is unexplained by differences in worker characteristics. For Hispanic women, 10 percentage points of the –15.9 percent difference in compensation compared with White men is accounted for by observable group differences; for example, differences in age, tenure, or location compared with White men. Hispanic women go from being the lowest compensated group to the third to lowest, being surpassed by both Black men and Black women.

In contrast to findings for Hispanic workers, observable differences in characteristics among Black male and female workers and White male workers explain very few of the differences in

---

[11] An estimate being statistically significantly different from zero is a measure of how confident we are that the underlying truth is not actually zero. Any estimate is subject to some variability because of the sample of data used. For example, the difference in average earning of White women compared with White men being statistically significantly different from zero means that we are confident that the true difference in average earnings is not zero. If an estimate is not statistically significantly different from zero, it means we cannot tell if the estimate is indicative of a true difference or is non-zero because of sampling variations and the underlying truth is zero. In the case of administrative data, we have the current population of DoD STEM workers. On the one hand, you could consider the current DoD STEM workforce as the only "true" population of DoD STEM workers, so that any observed difference would be thought of as the true difference with complete certainty. On the other hand, you could consider the current population of DoD STEM workers as one possible realization of all the people who could have possibly become DoD STEM workers. To be conservative, we follow the second notion and present statistical significance where relevant.

[12] Values from tables have been rounded in the text.

compensation. For Black women, things look even worse once observables are included. Looking at Table 2.2, the inclusion of age, length of service, and education exacerbates rather than explains the compensation difference between Black women and White men. This indicates that using the different age, length of service, and education profiles of Black women and White men, we would have expected Black women to earn more than White men. Instead of explaining why we observe White men earning more than Black women, accounting for the differences in age, length of service, and education make the pay gap even larger.

Table 2.2. Select Results from Regressions of Compensation on Department of Defense STEM Worker Characteristics

Demographic Group	(1)	(2)	(3)	(4)	(5)	(6)	(7)
White women	-6,210.0*** (-38.85)	-6,295.8*** (-49.67)	-6,292.4*** (-52.05)	-4,071.5*** (-33.38)	-6,361.8*** (-52.76)	-4,287.3*** (-35.25)	-3,696.7*** (-31.33)
Hispanic men	-7,648.8*** (-31.20)	-4,160.3*** (-21.20)	-3,499.4*** (-18.21)	-3,865.7*** (-20.25)	-3,477.6*** (-18.20)	-3,894.9*** (-20.52)	-3,513.4*** (-19.02)
Hispanic women	-13,967.8*** (-33.44)	-10,469.6*** (-30.41)	-9,156.7*** (-27.83)	-7,253.2*** (-22.57)	-8,963.1*** (-27.40)	-7,236.1*** (-22.61)	-6,509.4*** (-20.74)
Black men	-7,799.2*** (-36.63)	-6,571.2*** (-36.43)	-7,872.8*** (-45.35)	-7,843.1*** (-45.47)	-7,878.2*** (-45.64)	-7,795.5*** (-45.46)	-6,474.0*** (-38.95)
Black women	-9,456.4*** (-32.28)	-13,522.6*** (-53.39)	-14,805.8*** (-61.43)	-12,621.9*** (-53.29)	-14,895.4*** (-62.18)	-12,789.1*** (-54.19)	-10,787.5*** (-47.09)
Asian men	4,097.6*** (18.60)	1,855.2*** (11.05)	-1,935.2*** (-11.41)	-2,760.5*** (-16.32)	-2,107.8*** (-12.43)	-2,818.3*** (-16.68)	-2,807.5*** (-17.05)
Asian women	624.6 (1.63)	-923.2** (-3.15)	-4,814.1*** (-16.74)	-4,376.3*** (-15.44)	-4,940.2*** (-17.16)	-4,457.6*** (-15.73)	-4,323.2*** (-15.69)
Controls							
Pay plan							X
STEM category						X	X
Agency					X	X	X
Location			X	X	X	X	X
Demographics		X	X	X	X	X	X
Year	X	X	X	X	X	X	X
Observations	428,968	428,968	428,968	428,968	428,968	428,968	428,968

SOURCE: Authors' analysis of DMDC data on DoD civilian STEM workers (FYs 2018–2020) provided to RAND.
NOTE: Column (1) is from a regression that includes only year controls, which are the raw results from Figure 2.2. The fully adjusted results from Figure 2.2 correspond to column (7). Demographics include age, length of service, and educational attainment. T-statistics are presented in parentheses under the coefficient estimates. Stars indicate the significance level using a test that the true coefficient is equal to 0, which would indicate that there is no average compensation difference between the group and White men.
** $p < 0.01$, *** $p < 0.001$.

Decompositions of Compensation Differences by Demographic Group

The regression analysis indicates there are persistent pay gaps between White men and all other demographic groups. Once differences in observable characteristics are accounted for, White men have average wages that are significantly higher than those of each other demographic group. Regression analysis accounts for differences in the levels of observable characteristics—for instance, differences in education levels between White men and Black men. The regression analysis, however, does not explore possible differences within the same level of an observable characteristic. For example, the model adjusts for differences in education levels but does not allow for the possibility that different groups are paid differently for the same education level.

In this section, we decompose the raw difference in average wages into a part explained by differences in levels of observable characteristics (the difference between the "Raw" and "Adjusted" wages is displayed in Figure 2.2), a part explained by differences in returns to observable characteristics (*endowments*), and a part we cannot explain with the observable characteristics. Levels of observable characteristics are called *endowments*, and how groups are paid for the same endowments is called the *return* to an endowment. The data cannot explain why there are different returns to the same observable characteristic across groups, nor do they show what drives the unexplained differences. Understanding if and where these differences exist in the data is important for informing future research and policy recommendations.

We use Oaxaca-Blinder-Kitagawa decompositions (Jann, 2008) separately for each demographic group using White men as the reference group. The endowments and returns of White men are taken as the baseline.[13] Decomposition tells us how we would expect the average wages of a particular demographic group to change if they looked like (had the same endowments) or were paid like (had the same returns to endowments) White men. The decomposition also estimates the difference in average wages that is not explained by the observable characteristics, whether endowments or returns. The observable characteristics are the same as in the previous section. We include STEM worker characteristics and other employment characteristics (i.e., education level, age, tenure, job characteristics, location, agency, STEM category, and pay plan).[14] In addition to the overall decomposition, in Appendix B we look at which of the endowments or returns drive differences in compensation between a particular demographic group and White men.

---

[13] Recall that endowments refer to levels of the characteristics, such as education levels, while returns refers to how those levels are compensated, such as the rate of pay for a given education level.

[14] Pay plan is somewhat different from other job characteristics: Although workers presumably choose their location, agency, and STEM category, they likely do not choose their pay plan. In this section, we provide evidence of whether there are differences in pay plans or differences in returns to the same pay plans for different demographic groups. In Chapter 3, we return to the issue of differential use of pay plans.

Figure 2.3 presents the raw difference in compensation between each demographic group and White men as the baseline group: Values are interpreted as the difference between their average earnings, and positive values indicate that White men earn more. The differences are similar to the raw differences of Figure 2.1 and column (1) of Table 2.2.

Figure 2.3. Raw Difference in Compensation Between All Other Demographic Groups and White Men in the Department of Defense Civilian STEM Workforce

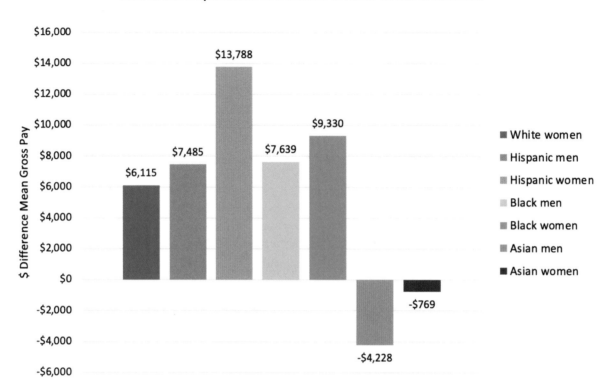

SOURCE: Authors' analysis of DMDC data on DoD civilian STEM workers (FYs 2018–2020) provided to RAND.
NOTE: The dollar value for each group is interpreted as the difference between the average compensation level of White men and that group's average compensation. A positive value indicates that White men have higher average compensation levels than that group. For example, White men earn on average $6,115 more annually than White women.

Figure 2.4 provides the endowments part of the decomposition. This is the difference in compensation compared with White men that is explained by differences in the levels of the characteristics, such as age and tenure.[15] Values are interpreted as how the average compensation of a demographic group would change if the group had the endowments of White men. For example, White women would earn $2,246 more if they had the endowments of their White male counterparts. As we show in Chapter 1 (Figure 1.4), White women are less likely to be in engineering and more likely to be in life sciences or social sciences than White men. Having the

---

[15] This piece of the decomposition is similar to the difference between the regression adjusted pay gap and the raw pay gap in Figure 2.2, which in turn is the difference between column (7) and column (1) in Table 2.2.

same endowments as White men—in this case, STEM category—would mean that White women were instead distributed across the STEM categories the same way as White men. Appendix B provides details on which endowments contribute to this difference. Asian men and women have endowments associated with higher compensation than the endowments of White men. White women, Hispanic women, Hispanic men, and Black men have endowments associated with lower levels of compensation than the endowments of White men, though noticeably less so for Black men than the other three groups. Differences in endowments between Black women and White men do not explain the observed difference in compensation.

Figure 2.4. Difference in Endowments Between All Other Demographic Groups and White Men in the Department of Defense Civilian STEM Workforce

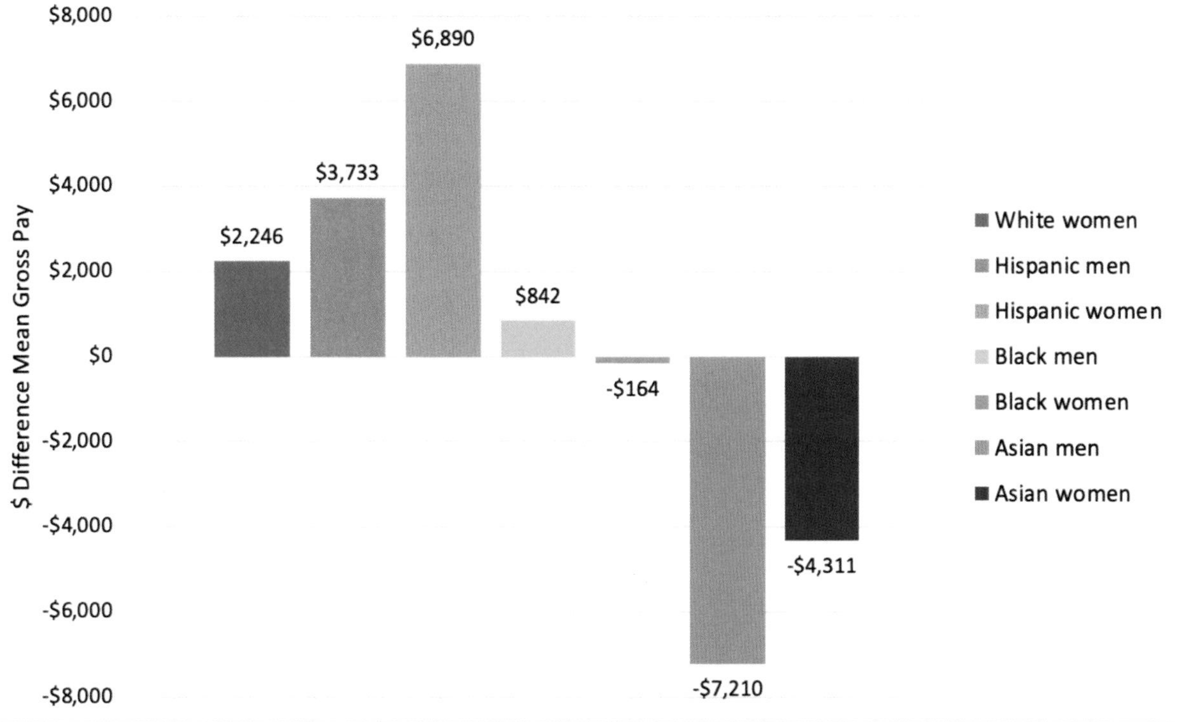

SOURCE: Authors' analysis of DMDC data on DoD civilian STEM workers (FYs 2018–2020) provided to RAND.
NOTE: In this figure, the dollar value represents how that group's average annual compensation would change if the group had the same endowments as White men. For example, White women are less likely to be in engineering and more likely to be in life sciences or social sciences than White men (Figure 1.4). These differences in endowments—the split across the STEM categories in this example—are associated with a difference in compensation because different STEM categories have different levels of average compensation. This chart shows the part of the total difference in compensation attributable to the differences in endowments. If White women instead looked like White men in terms of both worker and job characteristics, we would expect White women to earn $2,246 more in average annual compensation.

Figure 2.5 provides the returns to endowments part of the decomposition—i.e., the pay rates for given characteristics. This piece has no counterpart in the regression analysis of the previous section of this chapter. The regression analysis uses a constant rate of pay for the same characteristic across all groups. Differences in returns indicates that different groups are compensated at different rates for the same endowments. For example, White women and White men have a different distribution across the STEM categories—this is a difference in endowments. A difference in returns to endowments means White women and White men are paid differently within the same STEM category. It is important to note that these differences in rates of pay per observable characteristic, such as tenure, are limited to the observable worker and job characteristics we consider. White women, Hispanic men, Hispanic women, and Asian women have relatively higher pay rates than White men per the observable characteristics in our models. If these groups were paid the same as White men are paid, White women would earn $2,229 less, Hispanic women would earn $5,087 less, and Asian women would earn $2,244 less. For these groups, differences in returns compared with White men do not explain the observed pay gap because these groups are paid relatively better than White men. However, that is not the case for Black men and Black women, who have lower returns to endowments than do White men. The differences for Black women is noticeably severe: If Black women were paid like White men, we would expect Black women to earn over $7,500 more. Differences in returns to endowments may represent an attempt to compensate for the lower overall levels of compensation between groups, although this does not hold true for Black workers.

Figure 2.5. Difference in Returns to Endowments Between All Other Demographic Groups and White Men in the Department of Defense Civilian STEM Workforce

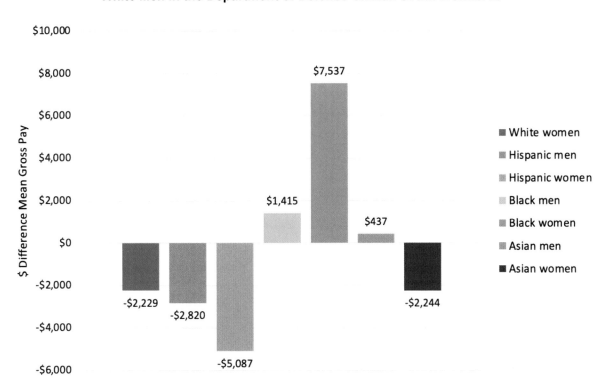

SOURCE: Authors' analysis of DMDC data on DoD civilian STEM workers (FYs 2018–2020) provided to RAND.
NOTE: The dollar value for each group represents how that group's average compensation would change if that group were paid the same way as White men for the worker and job characteristics included. For example, if White women were paid the same for the same levels of age, tenure, STEM category, location, agency, and pay plan, we would expect the average annual compensation of White women to decrease by $2,229. This indicates that although the overall pay gap between White men and White women is positive, differences in rates of pay for the worker and job characteristics included do not explain that pay gap.

This final piece of the decomposition is an unexplained difference, which is the disparity in pay when all other demographic groups are compared with White men, shown in Figure 2.6. This gap is smallest for Asian men and Black women.

Taken as a whole, the three pieces of the decomposition shed light on where the overall difference in compensation between a given demographic group and White men may come from. For Asian men, the difference in endowments accounts for most of the variance, with Asian men having higher education and working at higher paying location (see Appendix B). For Black women, the difference in returns to endowments drives most of the compensation disparity, which is in turn driven by different returns to age and tenure (see Appendix B). The unexplained gap compared with White men is largest for Hispanic women and Asian women. For all groups—but Hispanic women in particular—there are unobserved factors that are associated with White men having higher levels of compensation. It is important to note that we cannot explain differences in pay rates, differences in returns, or unexplained differences. In contrast to other demographic groups, Black women are paid relatively worse by age and tenure than White men. Unlike education level, which has a clear link to differences in compensation, there is no straightforward reason why rates of pay would differ by demographic group. These noticeable differences in pay rates illustrate an avenue for further investigation, particularly for the case of Hispanic women.

Figure 2.6. Unexplained Difference Between All Other Demographic Groups and White Men in the Department of Defense Civilian STEM Workforce

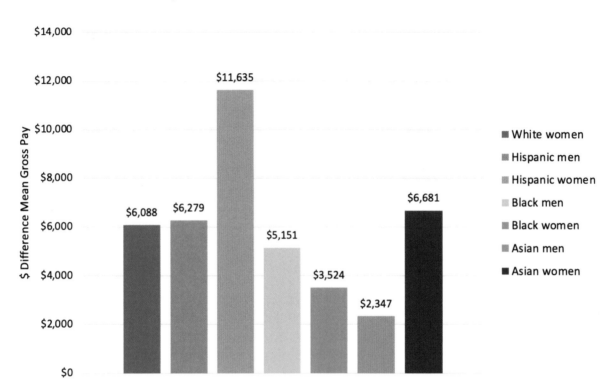

SOURCE: Authors' analysis of DMDC data on DoD civilian STEM workers (FYs 2018–2020) provided to RAND.
NOTE: The dollar value for each group represents the difference between the average annual compensation of White men and all other demographic groups that cannot be explained by worker and job characteristics. We can quantify an average pay gap by group that is not attributable to either differences in worker characteristics (endowments) or differences in how workers are paid for a given characteristic (returns); we cannot explain this gap with the variables in our model.

Summary

White male workers have higher rates of unadjusted compensation than all demographic groups except for Asian workers. However, when such characteristics as education, age, tenure, and location are included in our models, White men are compensated better than all groups. Within each racial and ethnic grouping, women are paid less than their male counterparts.

Although controlling for these characteristics generally closes the compensation disparity, it is exacerbated when Black women are compared with White men. Furthermore, our decomposition analysis indicates that the compensation rates for Black workers are lower than for White men. If Black women were compensated for their characteristics in the same manner as White men, they would receive an additional $7,500 annually on average.

There are significant compensation differences between White men and other demographic groups that cannot be explained by either differences in characteristics or differences in pay rates. Although these unexplained differences are relatively small for Asian males, the disparity for Hispanic women, at nearly $12,000, is almost double the difference for the next groups: Asian women (almost $7,000), White women (over $6,000), and Hispanic men (almost $6,300). These unexplained differences could be attributable to discrimination.[16] However, additional analysis and data with other measurable information (e.g., STEM worker performance levels) would be required to further reduce the level of uncertainty.

---

[16] For a discussion on interpreting the role of discrimination in gender pay gap analyses, see Blau and Kahn, 2017. For a similar discussion of pay gap analysis focused on race, see Kim, 2010.

Chapter 3. Deeper Dive into Pay Plans

The previous analysis indicates that pay plans matter for compensation and that pay plans have different racial and gender compositions. This finding is not surprising given all the ways in which pay plans vary, such as structure (e.g., narrower paygrades in GG and GS versus wider pay bands in demo plans), authorities (e.g., demo plans usually have direct hiring authorities), management practices (e.g., pay-for-performance bonuses in demo plans), and occupational categories (e.g., engineering sciences more highly represented than other STEM categories in demo plans). From the pay differential decompositions in Chapter 2 and Appendix B, differential representation in pay plans is a significant contributor to the pay gap between a demographic group and White men for five of the seven groups.[17] The results in columns (6) and (7) from Table 2.2 suggest that pay plan composition is a substantial factor in pay differentials on top of differences in education, location, STEM category and other observables. Column (6) accounts for education, age, length of service, location, STEM category and agency. Column (7) additionally controls for different pay plans. This accounts for groups being differentially represented across pay plans. The pay gap associated with a demographic group can then be understood as the result of comparing individuals within the same pay plan. The compensation gap for each group decreases from column (6) to column (7), indicating that compensation gaps are partly explained by differential representation across pay plans.

In this chapter, we further delve into the demographic composition of different pay plans commonly used in the DoD STEM workforce (see Table 1.1 in Chapter 1). We focus on six pay plans that fall into three categories:[18]

- GG and GS pay plans pooled together
- Lab Demo pay plans that are commonly used for DoD STEM workers and represent different DoD employers, specifically the Army (DB), Navy (DP), and Air Force (DR)
- the AcqDemo pay plan commonly used in the DoD STEM workforce (NH).

First, we examine differences in average compensation by pay plans. Next, we consider each pay plan's demographic composition. Finally, we look at differences in the highest pay grades

---

[17] In contrast, differential representation in pay plans shows that Asian men would be better compensated than White men. Asian women are the only group whose representation in pay plans is not significantly different than that of White men.

[18] The pay plans were grouped in this manner to put plans with similar characteristics together. GS and GG are similar; GG supports the intelligence community. DB, DP, and DR are all Lab Demo plans across the services. AcqDemo is a DoD-wide plan for acquisition professionals.

within pay plans.[19] Combined, these analyses demonstrate the role of pay plans in compensation differences between demographic groups in the DoD civilian STEM workforce.

We focus our pay plan analyses on the period of FYs 2016–2020 because 2020 is the last year pay period for which we have complete data, and the years prior to 2016 had high fluctuations in the number of workers within each pay plan of interest.[20] Table 3.1 provides the average compensation level within each of these pay plans over time, adjusted to FY 2021 dollars. Our measure of compensation is the same as in the previous chapter and throughout the report: gross pay. In each year, average compensation is lowest for the GG and GS pay plans and highest for the DB pay plan, which the Army uses for scientists and engineers.[21]

Table 3.1. Average Adjusted Compensation Levels Within Each Pay Plan (U.S. Dollars)

Pay Plan	Fiscal Year				
	2016	2017	2018	2019	2020
DB	30,281	134,317	128,422	127,123	130,098
DP	115,984	120,036	109,951	113,349	115,227
DR	126,113	130,218	125,367	123,190	125,690
GG/GS	97,127	101,197	95,441	95,062	95,962
NH	113,612	119,511	113,754	13,173	114,705
GG/GS	97,127	101,197	95,441	95,062	95,962

SOURCE: Authors' analysis of DMDC data on DoD civilian STEM workers (FYs 2016–2020) provided to RAND.
NOTE: Values are inflation-adjusted to FY 2021 dollars.

Demographic-Group Composition by Pay Plans

Figure 1.5 showed that White men are the most likely demographic group to be in pay plans other than GG and GS. Figure 3.1 shows the demographic makeup of each of the five pay plans considered. Consistent with the observation in Chapter 1, White men are relatively under-represented in the GG and GS plans in favor of the other plans, which in turn have higher average compensation rates. Notice that White men are least represented in the GG and GS pay plans (they make up about 60 percent of the STEM workers on a GG or GS pay plan) and most

---

[19] Demo pay plans, such as the Lab Demo and AcqDemo plans, do not have pay grades per se. Instead, they have pay bands that are much wider than in the GG and GS grades. However, for simplicity in presentation, we use the term *pay grade* to indicate a different level within a pay plan.

[20] Between 2010 and 2015, the number of DoD civilian STEM workers varied drastically in grades similar to GG, DP and NH. The number of civilian STEM workers with an NH pay plan doubled between 2015 and 2016, whereas the number of participants in GG and DP went from a handful in 2010 to several thousand in 2015. We expand from the 2018 cutoff of the compensation analysis back to 2016 to increase the chance of seeing promotions. Thus, the 2016 starting point is a balance between having enough years to see advancements and not including years of erratic movements across pay plans.

[21] Notice there is a pay bump in FY 2017. Although we do not know for certain what caused this, in combination with the federal civilian pay raise instituted by the Barack Obama administration, FY 2017 had an extra pay period. The subsequent dip in FY 2018 could be a result of cuts to cost-of-living adjustments under the Donald Trump administration's budget.

represented in the DR pay plan (they make up almost 75 percent of the STEM workers on a DR pay plan). Table 3.1 shows that the GG and GS pay plans have the lowest average compensation levels and the DB pay plan has the highest.

Figure 3.1. Demographic Composition of the Department of Defense Civilian STEM Workforce Across Pay Plans

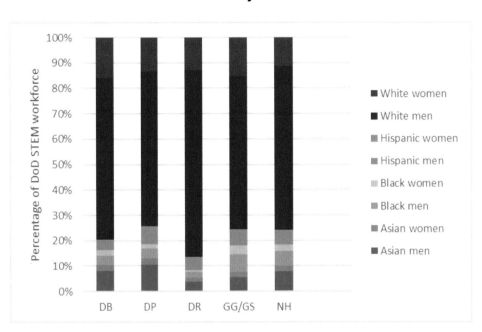

SOURCE: Authors' analysis of DMDC data on DoD civilian STEM workers (FYs 2016–2020) provided to RAND.

A key question is whether the relative overrepresentation of White men in alternative pay plans to GG and GS can be attributed to differences in qualifications. For example, perhaps White men are more likely to work in STEM categories that use alternative pay plans; in this case, the difference observed for White men would not be because of their race and ethnicity but because of the STEM category. To explore this possibility, we split the pay plans between GG and GS, and all other pay plans and use regression analysis to predict pay plan using demographic group and other observable characteristics (STEM category, agency, location, length of service, education, and age). The results are presented in Table 3.2 and Appendix C. The "Raw Probability" column in Table 3.2 lists the probability of being on an alternative pay plan to GG and GS for each group. This raw probability is the proportion of each group that is not on a GG or GS pay plan. The second column is regression-adjusted. The other worker and job characteristics are included along with demographic group identifiers in a regression

29

model.[22] The adjusted probabilities are closer across groups than the raw probabilities, indicating that adjusting for observable characteristics explains some of the differential use of alternative pay plans by demographic group. After adjusting for observable differences, Black men are almost 6 percentage points less likely to be on a pay plan other than GG or GS than White men, and Black women are over 7 percentage points less likely. These differences cannot be explained by age, tenure, education levels, STEM category, agency, or location. Future work that looks to understand the different use of pay plans using gender, race, and ethnicity will need to look beyond these worker and job characteristics.[23]

Table 3.2. Probability of Being on an Alternative Pay Plan by Demographic Group

Demographic Group	Raw Probability (%)	Adjusted Probability (%)	Difference (adjusted) Compared with White Men
White men	22.45	22.11	
White women	18.76	21.16	−0.96***
Hispanic men	19.81	20.91	−1.21**
Hispanic women	16.10	19.34	−2.77***
Black men	14.32	16.29	−5.82***
Black women	13.80	15.02	−7.09***
Asian men	27.86	22.09	−0.02
Asian women	25.12	22.41	0.30

SOURCE: Authors' analysis of DMDC data on DoD civilian STEM workers (FYs 2016–2020) provided to RAND.
NOTE: ** $p < 0.01$ *** $p < 0.001$. Outcome is a binary indicator for being on alternative pay plan. Logistic regression used and results are marginal predictive probabilities (adjusted probability) and average marginal effects (last column). Raw probability is the proportion of each group who are on alternative pay plans by group. For example, 22.45 percent of White men are not on GG or GS pay plans. The adjusted probability adjusts for the difference in alternative pay plans use by worker and job characteristic. The main variable of interest is the difference. For example, once differences in worker (aside from demographic group) and job characteristics are accounted for, Black women are 7 percentage points less likely to be on an alternative pay plan that White men.

Demographic Group Composition Within Pay Grades by Pay Plan

As a final analysis of pay plans, we look at the different pay grades within each pay plan. In this subsection, we consider the GG and GS plans separately. The full composition of the DoD STEM workforce across the different pay grades within each pay plan is given in Table C.2. We

---

[22] Logistic regression is used because it is a regression model designed for a binary outcome variable. The predicted outcomes from the regression model are understood as predicted probabilities that the binary variable takes as the value 1, which in this case means the probability that someone is not on a GG or GS pay plan. See Appendix C for more details.

[23] There may be many other job or worker characteristics that are correlated with gender, race, and ethnicity and are related to pay plan assignment, such as occupation. In the next chapter, we take a brief look at occupation classification within the GG and GS pay plans as it relates to compensation.

limit our analysis to pay grades with a sufficient number of STEM workers, as described in Appendix C.

Our final analysis looks at demographic composition in the top-paying grade of each pay plan, but limits this analysis to GG, GS, and NH. For the Lab Demo pay plans (DB, DP, and DR), the top grades did not have enough workers to provide a statistically accurate interpretation. The demographic composition of the top pay grades for GG, GS, and NH (i.e., GG-15, GS-15, and NH-04) is listed in Table 3.3. Although White men make up just over 61 percent of the DoD STEM workforce, they make up close to 70 percent of the top pay grades. No other demographic group is overrepresented in the top grades in these three pay plans. The small sample sizes of other demographic groups in each top pay grade limit our confidence in delving into these differences with any modeling techniques, but these summary statistics support the hypothesis that one reason White men do not have the highest levels of advancement is that they are more likely to have already hit the top pay grade. (See Appendix A for rates of advancement by demographic group within pay plan.)

Table 3.3. Demographic Composition of Top Pay Grades for Select Pay Plans

Demographic Group	Total Percentage	GG-15		GS-15		NH-04	
		N	Percentage	N	Percentage	N	Percentage
White men	61.28	541	72.33	5,655	69.32	6,542	69.71
White women	14.81	83	11.1	1,220	14.95	1,227	13.08
Hispanic men	4.61	25	3.34	175	2.15	331	3.53
Hispanic women	1.48	12	1.60	35	0.43	82	0.87
Black men	6.49	41	5.48	452	5.54	473	5.04
Black women	3.26	16	2.14	176	2.16	190	2.02
Asian men	6.17	21	2.81	327	4.01	419	4.47
Asian women	1.91	9	1.20	118	1.45	120	1.28
Total		748	100.00	8,158	100	9,384	100

SOURCE: Authors' analysis of DMDC data on DoD civilian STEM workers (FYs 2016–2020) provided to RAND.
NOTE: Total percentage gives the demographic composition of the DoD STEM workforce limited to the six pay plans considered in this chapter (DB, DP, DR, GG, GS, and NH).

Summary

In analyzing the pay plans, we found the best compensated pay plans were Lab Demos (DB, DP, and DR), closely followed by the AcqDemo plan used DoD-wide. The GS and GG play plans had the lowest-compensated individuals. Examining the demographics of these pay plans shows that White men made up just over 60 percent of the STEM workforce but occupied about 70 percent of the most senior grades in the data considered here. Additionally, White men and Asian workers are more likely to be on pay plans other than GG and GS, while Black women are the least likely.

Chapter 4. Case Study of Engineering and Information Technology and Computer Science Occupations in the Navy

In this chapter, we dive deeper into demographic trends within two of the largest STEM occupational categories in the DoD STEM workforce: engineering sciences and IT and CS. Unlike the previous chapters, this chapter narrows its scope by exploring occupational differences while holding location, agency, and STEM category constant. This analysis offers a more nuanced understanding of compensation trends, given that civilian employment decisions tend to happen at the local level.

We conduct this analysis by choosing the location (U.S. state), agency, and STEM category combination with the largest share of workers from each demographic group other than from White men. We then compare the top five occupations for that group with the top five occupations of White men in that same location, agency, and STEM category combination. For this chapter, we use data from FYs 2016–2020, as in Chapter 3.

Table 4.1 shows the location, agency, and STEM category with the largest population for each demographic group, the number of STEM workers in that group in that combination, and the number of White male workers in that combination. For each group, the Navy is the agency.[24] Engineering is the STEM category component with the largest population combination for all demographic groups except Black women, who are more represented in IT and CS occupations. Hispanic workers and Asian men are the most prevalent demographic groups represented in California. Black workers and White women are the most prevalent demographic groups represented in Virginia. Asian female workers are the most represented demographic group in Hawaii.[25]

---

[24] A single agency (the Navy) is the source for the group comparisons presented in this chapter, which could limit the generalizability of the findings to other DoD components (e.g., Army). However, many of these findings align with the overall analysis of DoD STEM compensation presented in Chapter 2, which lends some support to the potential for these findings to be generalized across DoD components.

[25] The use of "most prevalent" and "most represented" refers to the component of the combination with the largest representation. It is possible that there is a larger population of Asian women, for example, in another state, but that the Asian women in that state are spread over multiple agencies or STEM categories so that no combination of the three has more than 256 Asian women.

Table 4.1. Location, Agency, and STEM Category with Most Workers by Demographic Group

Group	N	State	Agency	STEM Category	White Men (N)
White women	651	Va.	Navy	Engineering	4,488
Hispanic men	741	Calif.	Navy	Engineering	3,689
Hispanic women	174	Calif.	Navy	Engineering	3,689
Black men	580	Va.	Navy	Engineering	4,488
Black women	277	Va.	Navy	IT and CS	1,701
Asian men	1,282	Calif.	Navy	Engineering	3,689
Asian women	256	Hawaii	Navy	Engineering	461

SOURCE: Author's analysis of DMDC data on DOD civilian STEM workers (FYs 2016–2020) provided to RAND.

Table 4.2 lists the average compensation (over FYs 2016–2020) for each demographic group in their respective combinations from Table 4.1 and the average compensation for White men in that same combination. Appendix C shows that the results from Table 4.2 hold true within each pay plan as well. Similar to what our analysis in Chapter 3 (Table 3.2) shows, White men are more likely to be on pay plans other than GG and GS than Hispanic or Black workers, not only on average but also for a particular location, agency, and STEM category.

Table 4.2. Average Compensation by Demographic Group for Location, Agency, and STEM Category

Demographic Group	Demographic Group (N)	Compensation ($)	Compensation for White Men ($)	Compensation Difference ($)
White women (Va.)	651	86,768	96,566	–9,798
Hispanic men (Calif.)	741	89,090	101,180	–12,090
Hispanic women (Calif.)	174	86,774	101,180	–14,407
Black men (Va.)	580	86,580	96,566	–9,986
Black women (Va.)	277	87,974	106,184	–18,210
Asian men (Calif.)	1,282	99,956	101,180	–1,224
Asian women (Hawaii)	256	99,475	97,708	1,767

SOURCE: Authors' analysis of DMDC data on DoD civilian STEM workers (FYs 2016–2020) provided to RAND.
NOTE: STEM categories are engineering for all demographic groups except Black women who work in the IT and CS fields.

Within location, agency, STEM category, and even pay plan, the compensation story is similar to that of the previous chapters. White men make more than most other demographic groups, and Asian workers tend to have the highest levels of compensation.

Diving Deeper Within a General Schedule Pay Grade and Single Occupational Category

Because of population size constraints with other pay plans, we focus our additional exploration of pay grade and occupation within the GG and GS pay plans. Within GG and GS paygrades, each demographic group is most represented at the GS-12 level, meaning that GS-12 has the largest percentage of workers per demographic group. Table 4.3 lists the average compensation as in Table 4.2, but the data are limited to GS-12 workers for each group.

Table 4.3. Average Compensation by Demographic Group for GS-12 STEM Workers, by Location, Agency, and STEM Category

Comparison Group	Comparison Group (N)	Compensation for Comparison Group ($)	Compensation for White Men ($)	Compensation Difference ($)
White women (VA)	268	85,811	95,786	−9,975
Hispanic men (CA)	280	99,709	100,430	−721
Hispanic women (CA)	63	94,824	100,430	−5,605
Black men (VA)	303	93,476	95,786	−2,311
Black women (VA)	121	81,529	86,835	−5,307
Asian men (CA)	523	102,084	100,430	1,654
Asian women (HI)	135	105,048	104,534	514

SOURCE: Authors' analysis of DMDC data on DoD civilian STEM workers (FYs 2016–2020) provided to RAND.
NOTE: The average compensation for each demographic group is for the location, agency, and STEM category listed in Table 4.1. For all demographic groups, the Navy is the agency. For all demographic groups except Black women, engineering is the STEM category. For Black women, the STEM category is IT and CS. The compensation for White men is limited to the same location, agency, and STEM category as the comparison demographic group.

Within GS-12, the pay gap between White men and most other groups is smaller than the overall gap: Asian men in California working in engineering for the Navy make more than corresponding White men. The gap between White women and White men in Virginia working in engineering for the Navy is almost unchanged. However, the pay gap among White men and Hispanic women and Black women remains substantial.

As a final exploration of observed pay gaps, we look at occupations. Despite being in the same location, agency, STEM category, and paygrade, occupation differences between the comparison groups could explain some of the pay gap in Table 4.3. Table 4.4 lists the top five engineering occupations for White women and White men in Virginia who work for the Navy. Table 4.5 provides the average compensation levels for White women and White men within the top occupation, engineering technician. We provide a similar presentation of results for Black women and White men in Tables 4.6 and 4.7, but instead focus on IT and CS occupations.[26]

---

[26] The population of Hispanic women in the most prevalent location, agency, and STEM category is too small ($N = 63$) to further divide by occupation. The resulting subsample would be too small to make any meaningful inferences for workers at large.

Table 4.4. Top Engineering Occupations for White Women and White Men Working for the Navy in Virginia on GG and GS Pay Plans

Demographic Group	N	Engineering Occupation
White women	146	Engineering technician (0802)
	131	Nuclear engineering (0840)
	79	Mechanical engineering (0830)
	58	Environmental engineering (0819)
	44	Civil engineering (0810)
White men	1,497	Engineering technician (0802)
	804	Nuclear engineering (0840)
	618	Mechanical engineering (0830)
	416	Electronics technician (0856)
	215	Electrical engineering (0850)

SOURCE: Authors' analysis of DMDC data on DoD civilian STEM workers (FYs 2016–2020) provided to RAND.
NOTES: Gray rows are for occupational categories that are common in both groups. The four-digit numbers in parentheses are the OPM occupational codes for these occupations. See OPM, 2018.

Table 4.5. Average Compensation Differences Between White Women and White Men in GS-12 Positions Within the Engineering Technician Occupation in the Navy in Virginia

Demographic Group	N	Compensation ($)
White women	70	89,951
White men	928	96,521
Difference		−6,570

SOURCE: Authors' analysis of DMDC data on DoD civilian STEM workers (FYs 2016–2020) provided to RAND.

Table 4.6. Top Information Technology and Computer Science Occupations for Black Women and White Men Working for the Navy in Virginia on GG and GS Pay Plans

Demographic Group	N	Engineering Occupation
Black women	242	Information technology management (2210)
	8	Operations research (1515)
	2	Computer science (1550)
White men	1,283	Information technology management (2210)
	193	Operations research (1515)
	22	Computer science (1550)

SOURCE: Authors' analysis of DMDC data on DoD civilian STEM workers (FYs 2016–2020) provided to RAND.
NOTES: The four-digit numbers in parentheses are the OPM occupational codes for these occupations. See OPM, 2018.

Table 4.7. Average Compensation Differences Between Black Women and White Men in GS-12 Positions Within the Information Technology Management Occupation in the Navy in Virginia

Demographic Group	N	Compensation ($)
Black women	119	81,781
White men	451	86,966
Difference		–5,185

SOURCE: Authors' analysis of DMDC data on DoD civilian STEM workers (FYs 2016–2020) provided to RAND.

Occupational differences do not explain the pay gap between White women and White men or Black women and White men. Limiting to the same location, agency, STEM category, occupation, and paygrade holds job characteristics constant. However, workers may differ in characteristics, such as education or age. To explore that possibility, we return to the regression results from Chapter 2. Using the full regression model, which includes both worker and job characteristics, we predict the average difference between each group using the difference in worker characteristics. The full regression gives estimated returns to age, education, and tenure, which can predict the expected difference in compensation between two workers with different ages, for example, but with all other characteristics being held the same. In this exercise, we predict the expected difference in compensation between White women and White men in Virginia working in engineering for the Navy and at the GS-12 level based solely on the differences in age, education level, and tenure in those two groups. We apply the same logic comparing Black women with White men for GS-12 jobs in IT and CS occupations in the Navy in Virginia.

Results of our analyses are presented in Table 4.8. Each column lists the predicted difference in average compensation between that group and White men using worker characteristics. For example, simply using the different age profiles of White women and White men who are engineering technicians in Virginia working for the Navy, we would expect those White women to make $2,124 less than White men. Summing across the characteristics provides the overall difference in compensation between that demographic and White men, e.g., using their qualifications and characteristics, Black women in IT management in Virginia are predicted to make over $4,500 more than White men in IT management in Virginia. Note that this exercise does not account for the observed compensation differences using demographic group, but instead looks solely at the predicted difference using job classification and other worker characteristics.

Table 4.8. Predicted Difference in Compensation of White and Black Women Compared with White Men Using Differences in Worker Characteristics

Group	Age	Tenure	Education	Total
White women	−2,124	1,048	227	−849
Black women	−1,264	2,523	3,418	4,677

SOURCE: Authors' analysis of DMDC data on DoD civilian STEM workers (FYs 2016–2020) provided to RAND.

NOTE: Group comparisons are with White men in the same occupation (for White women: engineering technician; for Black women: IT management), same pay plan and grade (GS-12), same employer (Navy), and same location (Virginia). Note that "Total" is the predicted difference in compensation solely using the listed characteristics and does not consider the pay gap associated with identification with a demographic-group.

Summary

Differences in worker characteristics do not explain the observed pay gaps between White women and White men in Navy GS-12 engineering technician jobs nor between Black women and White men in Navy GS-12 IT management jobs in Virginia. In fact, using worker characteristics, we would expect Black women to make over $4,500 more than corresponding White men, while they actually make an average of $5,185 less. We would expect White women to make $849 less than White men; in reality, White women make $6,570 less.

Chapter 5. Key Findings and Recommendations

Given Congress and DoD's continued interest in employing STEM workers and the federal government interest in promoting diversity, equity, inclusion, and accessibility, we explored factors that could explain compensation differences between demographic groups (gender and racial and ethnic) in DoD's civilian STEM workforce. We also looked at advancement and retention trends over time within this workforce, although limitations in the data precluded more detailed analyses of advancement outcomes.

In this chapter, we highlight the key findings of our deep dive into compensation differences among demographic groups within the DoD civilian STEM workforce. We conclude with recommendations for DoD policy regarding compensation of this important sector.

Key Findings

Our analyses produce several key findings about demographic-group differences in compensation within the DoD civilian STEM workforce.

White Men Have Higher Rates of Compensation Than Other Demographic Groups

When looking at raw rates of compensation, Asian men, Asian women, and White men are the highest-earning demographic groups and Hispanic women are the lowest-earning demographic group. These comparisons do not consider differences among workers other than demographic group. We consider worker characteristics (age, education level, and tenure) and job characteristics (agency, location, and STEM category) that could vary across demographic groups and are associated with different levels of compensation. Adjusting for differences in characteristics through regression analysis, White men have the highest levels of compensation and Black women have the lowest.

Women Earn Less Than Men Regardless of Observable Differences

Within race and ethnicity groupings, women earn less than men, both in terms of raw compensation and fully adjusted for observed differences. Fully adjusted for observed differences through linear regression, White men are the highest earners and Black women are the lowest. Asian women earn 1.8 percent less than Asian men, Hispanic women earn 3 percent less than Hispanic men, White women earn 4 percent less than White men, and Black women earn 4.4 percent less than Black men. These pay gaps are not explained by the observed characteristics.

Location Is Important for Asian STEM Workers

A large shift happens to compensation for Asian men and women when location is included in the controls. This indicates that differences in location between Asian workers and their White male counterparts explain why Asian workers receive higher compensation than White male workers; i.e., Asian workers tend to be in areas with higher average levels of compensation. Once workers in the same area are compared, Asian workers earn less on average than their White male counterparts by thousands of dollars.

The Same Characteristics Affect Demographic Groups Differently

In addition to differences in characteristics, we used decomposition analysis to look at differences in returns to characteristics. Although Asian women, Hispanic men, Hispanic women, and White women have lower (regression-adjusted) average compensation levels overall, they also have pay rates that are relatively favorable compared with those of White men using the observable characteristics in our models. For example, White women tend to have fewer years of service than White men, but White women are compensated for each additional year of service relatively better than White men. This could be because White women are earlier in their career trajectory and subject to higher raises. The data cannot tell us why these differences exist. By contrast, Black workers have lower returns to given characteristics than White men. The difference for Black women is noticeably severe. If Black women were paid like White men for the same qualifications, Black women would receive over $7,500 more in compensation. Although we can only speculate about why such differences exist, we can quantify the differences themselves and how the differences vary by demographic group.

There Are Substantial Compensation Differences Unexplained by Observable Characteristics

For both Hispanic male and female workers, differences in worker characteristics compared with White men explain most of the raw differences in compensation. For Hispanic men, observable differences explain over $4,100 of the $7,649 raw difference, leaving $3,513 unexplained by control variables. In the case of Hispanic women, observable differences account for 10 percentage points of the –15.9 percent difference in compensation compared with White men. Hispanic women go from being the lowest-compensated group to the third-lowest: Both Black men and Black women receive less pay.

For Black male and female workers, observable differences explain very little of the differences in compensation compared with White men. For Black women, observable differences actually exacerbate rather than explain the compensation difference.

White women, Hispanic women, Hispanic men, and Black men have endowments associated with lower levels of compensation than the endowments of White men, though noticeably less so for Black men than the other three groups. Differences in endowments between Black women and White men do not explain the observed difference in compensation.

Representation in Pay Plan Makes a Difference

Pay plans explain some of the pay gap between White men and other demographic groups. Demographic group identity, in turn, predicts pay plan on top of other observed characteristics. This means that although some of the compensation gap uses differences in pay plan, there are also unexplained differences in pay plans across demographic groups. White men are relatively overrepresented in better-compensated pay plans and within pay plans, White men are relatively overrepresented in the top pay grade.

After adjusting for observable differences, Black men are almost 6 percentage points less likely to be on a pay plan other than GG and GS, and Black women are over 7 percentage points less likely to be on another pay plan than White men.

Case Study Lessons: Controlling for STEM Category, Location, and Agency Still Does Not Fully Explain Compensation Differences

Within location, agency, STEM category, and even within pay plan, White men make more than most other demographic groups, and Asian workers tend to have the highest levels of compensation. Within pay grade GS-12, the pay gap between White men and most other groups is smaller than the overall gap. The gap between White women and White men working in engineering in Virginia for the Navy is almost unchanged when we further limit to the GS-12 pay grade, but the pay gaps between Hispanic women and White men or Black women and White men remain substantial. As a final exploration of observed pay gaps, we look at occupations.

Occupational differences within engineering and IT and CS do not explain the pay gap between White women and White men or Black women and White men. Differences in worker characteristics do not explain the observed pay gaps. In fact, using worker characteristics, we would expect Black women in the information technology management occupation within IT and CS in Virginia working for the Navy to make over $4,500 more than corresponding White men, whereas in the data, they make an average of $5,185 less. Using the differences in worker characteristics for White women and White men who are engineering technicians in Virginia working for the Navy, we would expect White women to make $849 less than White men. In reality, those White women make $6,570 less than the corresponding White men. Therefore, additional analyses could help further understand the unexplained differences.

Recommendations

Drawing on our findings, we offer three recommendations for DoD to better understand and address potential inequities in compensation for its civilian STEM workforce. For each recommendation, DoD should coordinate or at least communicate with OPM as the lead agency mandated by presidential Executive Order to review pay inequities as a DEIA barrier in the federal civilian workforce.[27]

Establish Additional Guidance for Department of Defense Component Use of Alternative Pay Plans for Department of Defense Civilian STEM Workers

Our analyses show that pay plans other than GG and GS, namely the Lab Demo and AcqDemo plans, are associated with higher compensation levels overall. We also show demographic distribution across pay plans varies: White men are overrepresented in these alternative plans relative to GG and GS once we account for other observable characteristics.

Although there is a process for organizations to apply for use of a demo plan, it is not clear how, at a component-wide level (e.g., Department of the Navy), an organization's use of a demo plan affects opportunities for STEM workers from various demographic groups.[28] For example, if non-White engineers are less likely to work in a military department's research laboratories but have similar worker characteristics (e.g., education levels) to the White engineers who work in those laboratories, what are the implications for compensation and other career outcomes for non-White engineers in that military department writ large? Previous RAND research highlights this concern: When comparing AcqDemo professionals with similar GS acquisition professionals in DoD organizations, Lewis et al. (2017) showed demographic-group differences in salary and promotion outcomes across the pay plan types.

To address this concern, DoD could establish additional guidance for DoD components (e.g., military departments, fourth estate agencies) about when and how to apply alternative pay plans, particularly demo plans. The goal of the guidance would be to ensure that DoD components benchmark compensation and other career outcomes for similarly qualified individuals across their organizations to ensure that components do not create inequalities in opportunities through the use of different pay plans. If there is evidence of potential inequalities across organizations within a component, the guidance could stipulate that be reported to the Office of the Secretary of Defense (OSD) (e.g., the Office of the Under Secretary of Defense for Research and Engineering, which has policy authority for the military laboratories).[29] This type of guidance

---

[27] Specifically, Executive Order 14035, Section 12, mandates that the Director of OPM "review Government-wide regulations and guidance . . . in order to address any pay inequities and advance equal pay" (Biden, 2021).

[28] As an example, in 2020, DoD submitted a notice in the Federal Register to apply a Lab Demo plan for the Army Research Institute for the Behavioral and Social Sciences (DoD, 2020). The notice refers to other organizations that want to use of alternative pay plans and draws comparisons with other military laboratories, but there is no assessment of the implications for using demo plans across the Army's entire science and technology population.

[29] The demo projects are limited to specific workforces (acquisition community, science and technical laboratories). Thus, even if there are comparable STEM workers elsewhere in a component, they might not work in an organization eligible for a demo pay plan. This emphasizes why DoD needs to understand potential gaps across organizations within components and determine if additional types of demo projects are warranted.

could be particularly timely given news in 2022 of plans to expand flexibilities within the Lab Demo plans (FEDweek, 2022).

Direct the Services to Assess STEM Compensation and Career Outcomes by Organization and Location to Better Address Unexplained Demographic-Group Differences

Although our case study sought to further explore compensation differences among demographic groups by holding the DoD component, geographic location, and STEM category constant, we were not able to fully explain the differences. Given that civilian personnel decisions tend to happen at a local level, addressing any potential inequities likely requires a local-level review of policies and practices. Edwards et al. (2021) advocate for the use of qualitative methods (e.g., surveys) to supplement quantitative methods, such as those used in this report. A survey, for example, could ask workers how they decided to apply for civilian positions and the paths they took to apply, and then link that information with entry-level administrative data (e.g., pay grade and starting salaries).[30] This could reveal how demographic groups take different pathways to enter an organization, and how those pathways affect the level at which they start in an organization. As we note in Chapter 1, previous RAND research shows that paygrade at entry has implications for women's ability to reach upper civilian grades (Keller et al., 2020).

It might be the case that internal audits happen within local organizations (e.g., within an Army command) but that information is not made available to OSD. We therefore suggest as a course of action that OSD direct the services to conduct assessments of organizations at specific locations to understand how local policies, practices, and labor market conditions affect demographic-group differences in compensation and other career outcomes. The services would then report on the results of their assessments. This course of action would follow the recommendation of Edwards et al. (2021) that DoD should not consider STEM workers as a monolith, given their wide variety of educational and occupational backgrounds.

Conduct Additional Analysis to Understand the Impact of Entry-Level Compensation on Demographic-Group Differences in Compensation

Given our resource and data limitations, we were not able to investigate how entry-level differences compensation among demographic group affected our findings. Previous analysis on Air Force civilians shows that demographic-group differences in entry-level paygrade can affect

---

[30] OPM's Federal Employee Viewpoint Survey, which fields U.S. federal civilian workers annually, includes questions about satisfaction with various workplace and job characteristics, including pay (OPM, undated-b). Other labor market surveys from outside organizations, such as Gartner or Gallup, may also provide insights into employee views on compensation and other factors affecting retention.

advancement to higher grades (Keller et al., 2020).[31] We suggest that future analysis on demographic-group differences in entry-level compensation include data on hiring bonuses, which we expect are applied differently across pay plans and STEM occupations. We would also suggest monitoring these differences over time, given the many changes in federal hiring trends in recent years and the U.S. economy writ large. Assessing changes over time and making comparisons with prior historical periods that were relatively stable could help policymakers understand whether recent trends reflect long-standing differences or continually evolving circumstances.

Conclusion

DoD remains interested in STEM disciplines and is examining its workforce through a DEIA lens. Available data indicate there are still unexplained differences in compensation among demographic groups. Although our findings advance this research, there is still work to be done. The proposed courses of action should, at a minimum, provide DoD increased understanding as it addresses DEIA issues in its STEM civilian workforce.

---

[31] Keller et al. (2020) analyzed demographic-group differences in entry-level paygrades for Air Force civilians. Although this analysis had a different focus, the methodological approach could be applied to analysis of compensation differences at entry.

Appendix A. Advancement and Retention Trends

In this appendix, we present average advancement rates and retention rates by demographic group.

Advancement

Because different pay plans have different grades and likely different underlying patterns of advancement, we consider the five pay plans separately. Figures A.1 through A.5 present the annual advancement rates by demographic group for each pay plan for FYs 2016–2020.[32] The vertical axis is the probability of *advancement*, defined as moving up a pay grade. This can also be understood as the proportion of the workers in that group who moved up a pay grade. One noticeable feature is that White men do not have the highest advancement rates within any of the pay plans considered. As we explore in the "Pay Grades" section, this is potentially because White men already are in higher pay grades and thus have less room for advancement.

The small numbers of non-White men in alternative pay plans limit our statistical analysis. Additionally, because of the small sample sizes, little changes can result in large fluctuations within a demographic (see Figures A.2 through A.5). Instead, the main text explores the different use of pay plans across demographic groups and the demographic representation across pay grades within the GG and GS pay plans.

---

[32] As we note in Chapter 3, we limit our analyses to FYs 2016–2020 because there were significant fluctuations in pay plan populations prior to this period. Also, 2020 is the last FY for which we had complete data.

Figure A.1. Annual Advancement Rates by Demographic Group for General Government and General Schedule Pay Plans

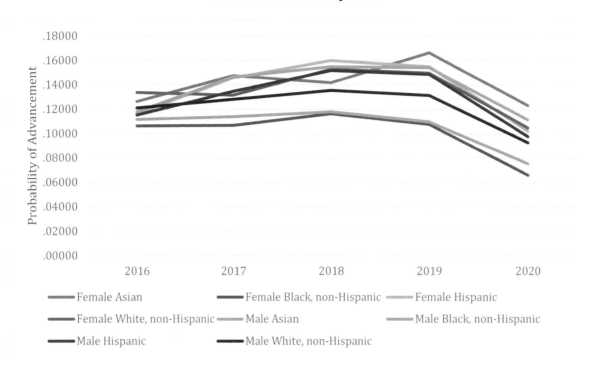

SOURCE: Authors' analysis of DMDC data on DoD civilian STEM workers (FYs 2016–2020) provided to RAND.

Figure A.2. Annual Advancement Rates by Demographic Group for Demonstration Engineers and Scientists Pay Plan in the Army

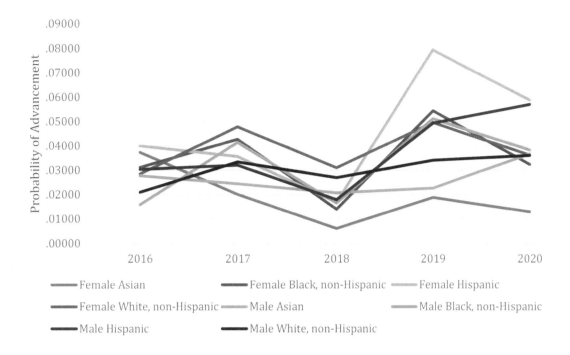

SOURCE: Authors' analysis of DMDC data on DoD civilian STEM workers (FYs 2016–2020) provided to RAND.

Figure A.3. Annual Advancement Rates by Demographic Group for Demonstration Professional Pay Plan in the Navy

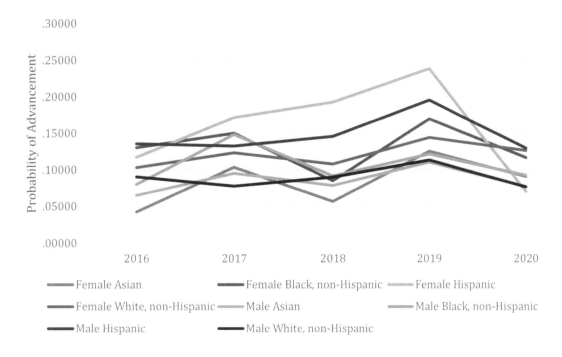

SOURCE: Authors' analysis of DMDC data on DoD civilian STEM workers (FYs 2016–2020) provided to RAND.

Figure A.4. Annual Advancement Rates by Demographic Group for Demonstration Professional Pay Plan in the Air Force

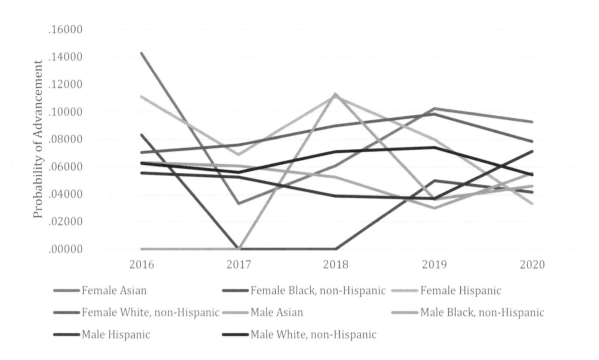

SOURCE: Authors' analysis of DMDC data on DoD civilian STEM workers (FYs 2016–2020) provided to RAND.

Figure A.5. Annual Advancement Rates by Demographic Group for Business Management and Technical Management Professional Pay Plan

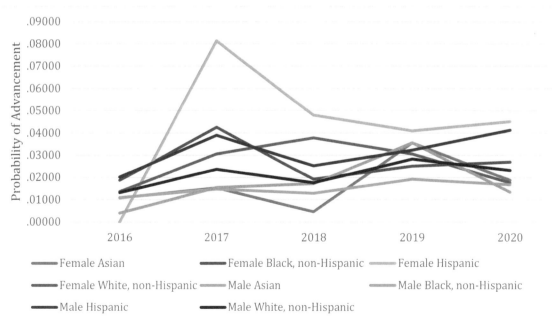

SOURCE: Authors' analysis of DMDC data on DoD civilian STEM workers (FYs 2016–2020) provided to RAND.

Retention

Figure A.6 presents average retention rates by demographic group over the 11 years from FYs 2010 to 2020. Rates of retention are high, and there are small differences by demographic group. Hispanic women generally have the lowest retention rate, although Asian men generally have the highest. This mirrors compensation rankings in which Hispanic women have the lowest raw average compensation levels and Asian workers have the highest. Focusing on FYs 2018–2020 as in the compensation analysis, the difference in retention rates between the highest group and lowest group is between 1.8 percentage points (FY 2020) and 3.2 percentage points (FY 2018). A retention rate gap of 3.2 percentage points is relatively small. From a baseline retention rate of 88 percent, as is observed for Hispanic women in 2018, a gap of 3.2 percentage points is a 3.6 percent difference in retention, meaning the retention highest retention rate in 2018 was 3.6 percent higher than the lowest retention rate. The retention gap fell to 1.8 percentage points in 2020, which is a two percent difference. Compared with the differences in average compensation (see Figure 2.1), which form a baseline of White male compensation rates ranging from +4.4 percent to –15.9 percent, the differences in retention rates are small. The large differences observed in compensation levels by demographic groups are not reflected in large differences in retention rates by demographic group.

Figure A.6. Retention Trends by Demographic Group

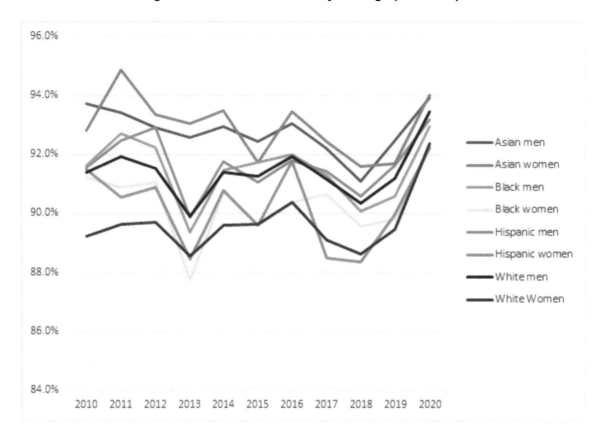

SOURCE: Authors' analysis of DMDC data on DoD civilian STEM workers (FYs 2010–2020) provided to RAND.

Appendix B. Detailed Decomposition Results for Compensation Analysis

We provide decomposition details for each of the seven non-White male demographic groups. We first consider White women compared with White men and walk through each aspect of the decomposition in detail. For the remaining six groups, we suppress some of the details; however, the same process was followed for Asian, Black, and Hispanic male and female groups. Please refer to the subsection for White women for more details regarding the interpretation of the decomposition results.

White Women

We first consider the compensation of White women compared with White men. Figure B.1 provides the overall decomposition of the difference in average compensation. The overall difference is listed first in the light green bar labeled "Difference." For White women, we observe an overall difference in average compensation compared with White men of just over $6,000. This means that White women make on average $6,000 less than White men. This is akin to the "Raw" difference in average compensation listed in Figure 2.2 and the first column of Table 2.2.[33] The following three bars in Figure B.1 are the results of the decomposition. The overall difference is the sum of the difference attributable to different endowments, the difference attributable to different returns to endowments, and the remaining unexplained difference.[34] The "Endowments" bar indicates the difference in average wages between White women and White men that are attributable to differences in endowments (worker characteristics). The endowments of White men are taken as the baseline. The "Returns" bar in Figure B.1 indicates the difference in average wages between White women and White men that are attributable to differences in returns to the same endowments. The "Endowments" bar accounts for differences in education (we saw in Chapter 1 women tend to have higher education levels than men). The "Returns" bar considers whether White women and White men with the same level of education are paid differently. The "Unexplained" bar represents the average difference in compensation between White women and White men that cannot be explained by either different levels of endowments or different returns to endowments.

---

[33] The actual numerical values are slightly different because Table 2.2 adjusts for overall average compensation changing by year, whereas the overall difference in the decomposition makes no adjustments.

[34] There is also an "interaction" piece of the decomposition that is a mathematical by-product of the other differences. It is generally small and often statistically insignificant. The interaction is difficult to interpret in context and is omitted from the results.

Figure B.1. Overall Decomposition of the Difference in Average Wages Between White Women and White Men

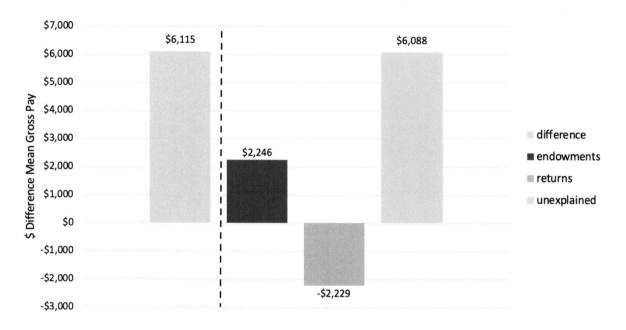

SOURCE: Authors' analysis of DMDC data on DoD civilian STEM workers (FYs 2018–2020) provided to RAND.

The $2,246 attributable to differences in endowments means we would expect the average wages of White women to be $2,246 higher if they had the same endowment levels as White men. This is akin to the difference between the "Raw" number and the "Adjustment" number from Figure 2.2, which is also the difference between column (1) and column (7) of Table 2.1. In other words, differences in endowments explain just over $2,000 of the observed pay gap between White women and White men. Panel B of Figure B.2 provides more details. This figure breaks down the total amount attributable to differences in endowments into the endowments that drive that difference. Only endowments that are attributable to statistically significant differences in compensation are included.[35] Education has a large, negative effect: Women have higher average levels of education than men, and we expect compensation to increase with education level. The education bar indicates that if White women had the same education levels as White men, we would expect White women to have lower wages by $2,282 on average (because White men on average have lower education levels than White women). White women have lower wages than White men already. If they also had the same education levels as those of White men, we would expect the average pay gap to increase by over $2,000.

---

[35] Recall that statistical significance means we have a high level of confidence (we use 95 percent throughout) that the value is not zero. A value of zero would imply no difference, so saying an endowment is statistically significant means there is only a small chance (five percent or less) that the observed value is really a noisy measure of a true zero. That chance is deemed small enough that we conclude that the true value is not zero so that the endowment truly does contribute to differences in compensation.

Panel B of Figure B.2 digs deeper into the "Endowments" and "Returns" bars from Figure B.1 and indicates that White women tend to work in higher paying agencies and higher paying locations than White men. This is indicated by the negative values for the "Agency" and "Location" bars. Instead of explaining the pay gap between White women and White men, differences in education level, agency, and location exacerbate it: If White women had the same education levels, agency of employment, and locations as White men, we would expect a larger pay gap than the one we observe. By contrast, differences in age, tenure, STEM category, and pay plan explain part of the observed pay gap between White women and White men. The $1,899 on the "STEM Category" bar indicates that White women work in lower-paying STEM categories than White men on average. If instead White women were distributed across STEM categories the same way that White men are, we would expect White women's wages to increase by $1,899. This means that differences in STEM category explain close to $2,000 of the overall pay gap between White women and White men. Differences in age, tenure, and pay plan also substantially contribute to the pay gap. Figure B.2 does not include unexplained differences.

Figure B.2. Detailed Decomposition of Average Wages Between White Women and White Men

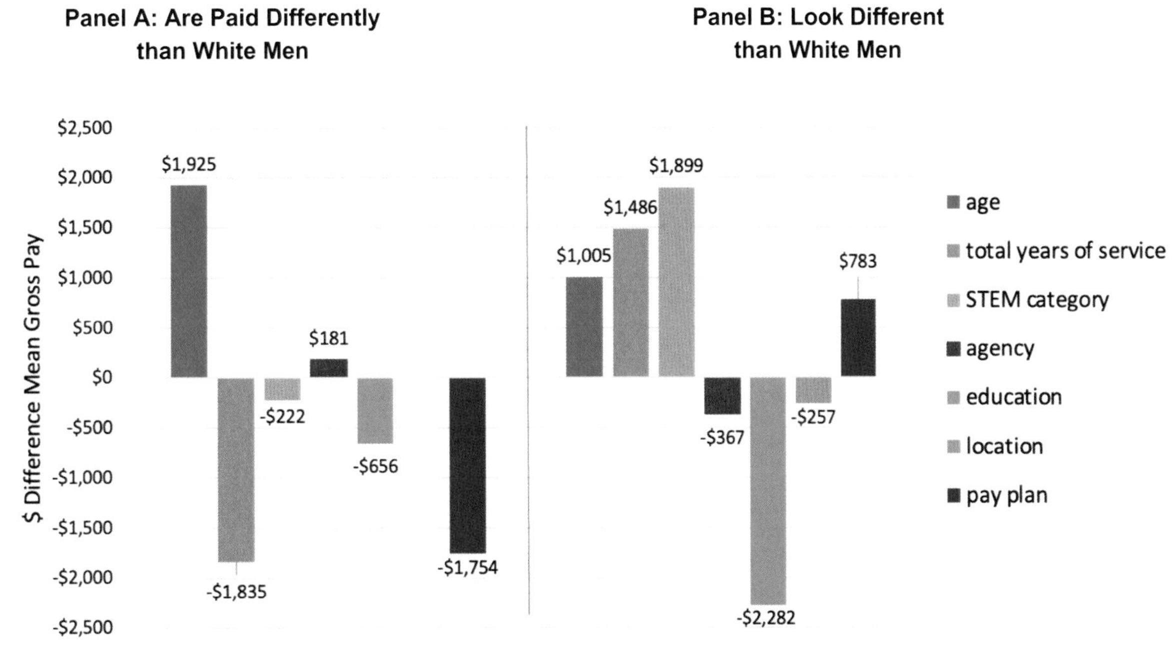

SOURCE: Authors' analysis of DMDC data on DoD civilian STEM workers (FYs 2018–2020) provided to RAND.

Looking at the "Returns" bar of Figure B.1, if White women were paid the same way as White men for their observable characteristics, we would expect the average compensation of White women to fall by $2,229. Although White women have observable characteristics that lead to lower pay than that of White men, White women tend to be paid better than White men with the same observable characteristics. Panel A of Figure B.2 provides more details. Like Panel B for endowments, Panel A lists the specific returns that are different between White women and White men. The $1,925 value for "Age" in panel A indicates that White women and White men of the same age are paid differently; White women are paid $1,925 less than White men of the same age. Conversely, White women are better compensated by tenure than White men. If instead, White women were paid for length of time in their jobs the way White men are paid, we would expect average wages for White women to fall by $1,835. Not only do White women tend to have higher levels of education than White men, White women are also better compensated by education level. If instead White women were paid like White men are for education, we would expect the average compensation of White women to fall by $656. Although White women tend to be in lower paying pay plans than White men, White women are on average better compensated for the same pay plan. We take a closer look at a subset of pay plans in Chapter 3.

Finally, we turn to the "Unexplained" bar in Figure B.1. For White women compared with White men, the differences in endowments and returns are roughly equal in magnitude in opposite directions. White women have endowments that are associated with lower levels of compensation than the endowments of White men, yet White women are on average better compensated for their endowments. Although we cannot say why White women have different returns than do White men, we can identify and quantify the extent of those differences. If the observable characteristics that we considered included all characteristics that determine compensation, we would expect White women and White men to make similar wages on average. Instead, there is a $6,088 pay gap that is attributable to unobserved differences between White women and White men. There are numerous factors not included in the observed characteristics that could contribute to this pay gap. Although we consider different pay plans, we do not consider different levels of seniority beyond accounting for differences in tenure. Occupation is not included. There are unobserved individual characteristics that could affect compensation, such as attachment to the labor force or value of outside working options. Although we include location, we have no data on individuals' marital status or family structure, both of which are generally thought to be correlated with attachment to the labor force, particularly for women. This unexplained difference could represent implicit or explicit discriminatory practices in compensating White women compared with White men, but such a conclusion cannot yet be reached with this information. We have listed numerous other unobserved factors that could lead to possibly large differences in compensation between these two groups.

Hispanic Men

Figure B.3 gives the overall decompensation results for Hispanic men compared with White men. The regression analysis indicates a large, raw pay gap of $7,649, which is to say that Hispanic men receive $7,649 less in average wages than White men. Accounting for differences in observed characteristics explains over half of that gap. The decomposition results mimic the insights from the regression analysis. The overall difference in average wages is estimated at $7,485. Roughly 50 percent—$3,733—of that difference is attributable to differences in observed characteristics. After accounting for observable differences in endowments, the regression analysis indicates a remaining pay gap of $3,513 between Hispanic men and White men (column 7 of Table 2.2). Using only the regression analysis, we might conclude that roughly half of the total difference is explained and the remaining $3,500 is unexplained, but the decomposition gives different conclusions. Although differences in endowments account for roughly half of the overall difference in compensation, Hispanic men are on average better paid for their endowments than are White men. If instead Hispanic men were paid the way White men are paid for the same observable characteristics, we would expect Hispanic men to make $2,820 less on average. This difference in the returns to endowments offsets much of the difference in compensation attributable to differences in endowments. The result is an estimated difference of $6,279 that is unexplained by either different endowments or different returns to endowments.

Figure B.3. Overall Decomposition of the Difference in Average Wages Between Hispanic Men and White Men

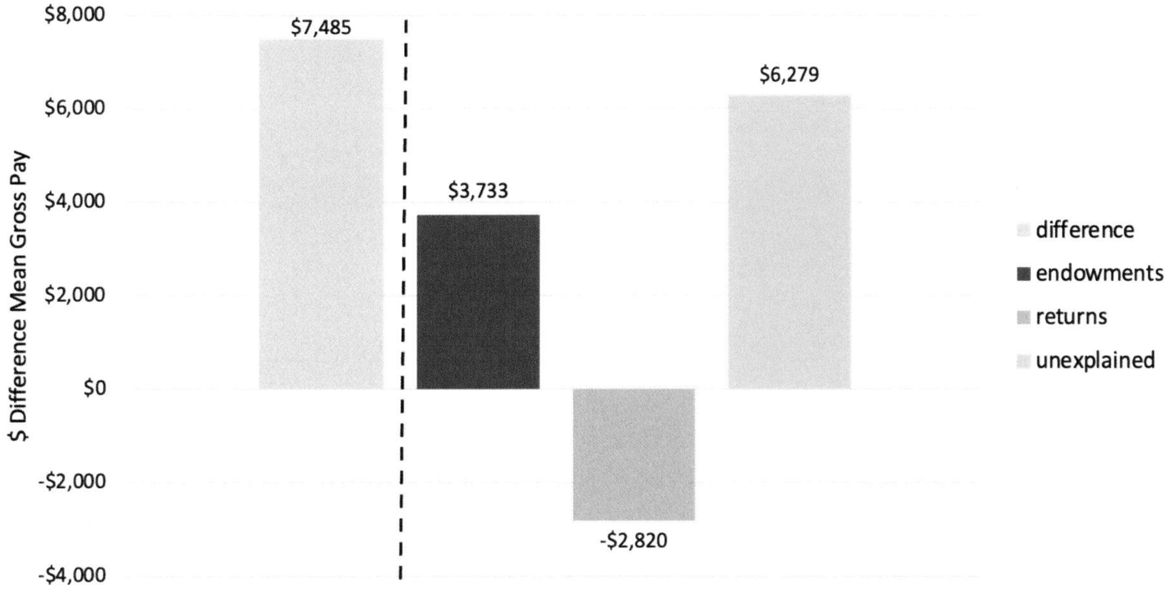

SOURCE: Authors' analysis of DMDC data on DoD civilian STEM workers (FYs 2018–2020) provided to RAND.

Figure B.4 provides more details on the differences in endowments (Panel B) and the differences in returns (Panel A) between Hispanic men and White men. Similar to White women, Hispanic men's differences in age and tenure compared with White men are associated with lower wages. If Hispanic men had ages and tenures similar to those of White men, we would expect their average compensation to increase by $1,173 and $1,739, respectively (or $2,912 total). Hispanic men tend to have lower levels of education than White men and work for lower-paying agencies but in higher-paying STEM categories. Hispanic men tend to be on lower-paying pay plans than White men but are better compensated according to pay plan.

Figure B.4. Detailed Decomposition of Average Wages Between Hispanic Men and White Men

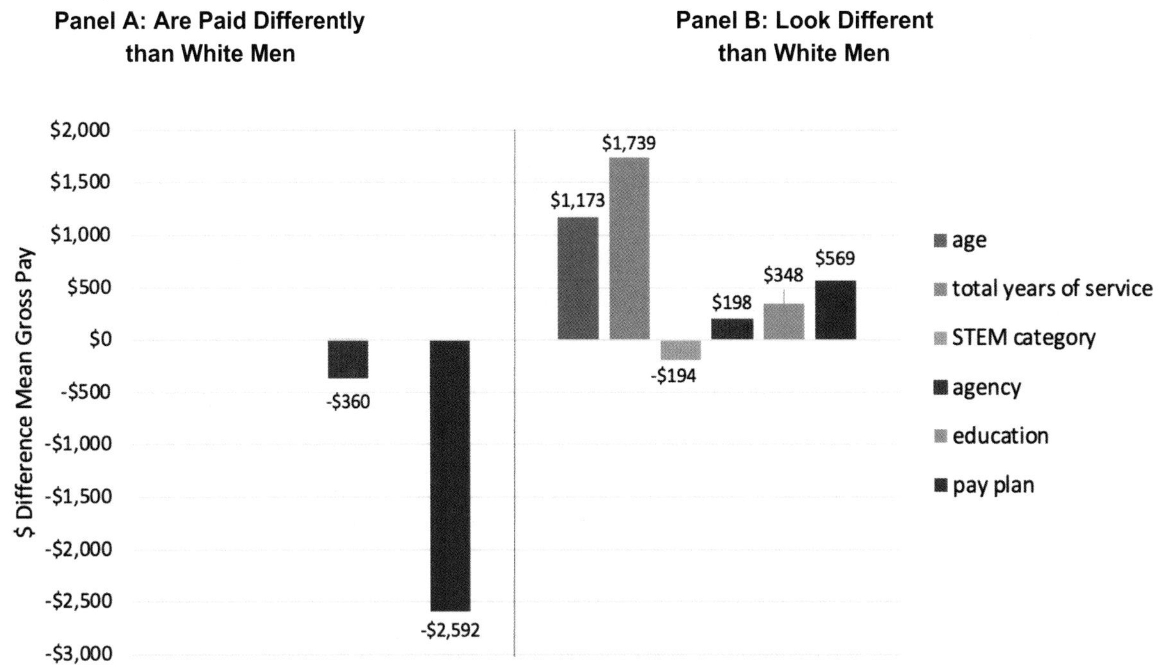

SOURCE: Authors' analysis of DMDC data on DoD civilian STEM workers (FYs 2018–2020) provided to RAND.

Hispanic Women

Hispanic women have the lowest average wages of any demographic group, with raw average wages almost $14,000 lower than those of White men. Differences in observable characteristics between Hispanic women and White men explain a large portion of those wages. From Figure 2.2 or Table 2.2 column (7) versus column (1), difference in observable characteristics explain 53 percent of the $14,000 raw pay gap; the resulting gap of $6,509 is unexplained by differences in observable characteristics. The overall decomposition results presented in Figure B.5 replicate those observations, but again reveal that differences in levels of observed characteristics are not the entire story. Hispanic women are paid relatively better than White men for the same characteristics. If Hispanic women were paid like White men, we would expect the average wages of Hispanic women to decrease by $5,087, further exacerbating the pay gap. Instead of an unexplained difference of roughly $6,500, accounting for both differences in levels of and returns to endowments leaves a resulting unexplained difference of $11,634.

Figure B.6 provides further details for differences attributable to endowments (Panel B) and returns to endowments (Panel A). Panel B shows that Hispanic women have higher average levels of education than White men, and would be paid $1,374 less, on average, if they had the same educational distribution as White men. On the one hand, all other differences in endowments work against Hispanic women. The age distribution and tenure of Hispanic women compared with White men results in lower wages for Hispanic women by $2,262 and $3,036, respectively. Hispanic women are more represented in lower-paying STEM categories and in lower-paying pay plans. On the other hand, Hispanic women are paid better than White men for many characteristics. Hispanic women have higher returns to tenure, STEM category, location, and pay plan.

Figure B.5. Overall Decomposition of the Difference in Average Wages Between Hispanic Women and White Men

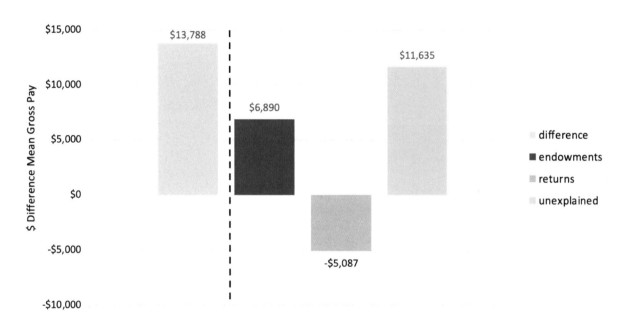

SOURCE: Authors' analysis of DMDC data on DoD civilian STEM workers (FYs 2018–2020) provided to RAND.

56

Figure B.6. Detailed Decomposition of Average Wages Between Hispanic Women and White Men

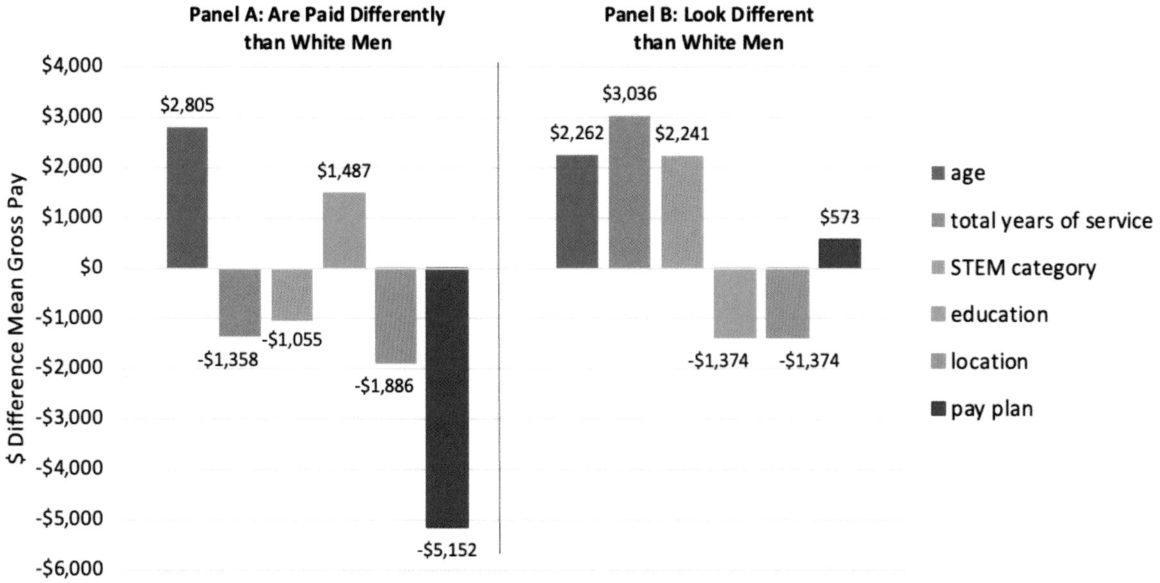

SOURCE: Authors' analysis of DMDC data on DoD civilian STEM workers (FYs 2018–2020) provided to RAND.

Black Men

Black men have similar raw compensation levels to Hispanic men: Both groups make close to $8,000 less than White men (Figure 2.2 and Table 2.2). Unlike Hispanic men, Hispanic women, or White women, however, differences in observable characteristics explain little of the difference in compensation between Black and White men. Figure B.7 gives the overall decomposition results for Black men compared with White men. Differences in endowments explain less than $1,000 of the over $7,500 difference in average compensation. Additionally— and again unlike Hispanic men, Hispanic women, and White women—Black male workers are paid worse than White men for the same endowments. If Black men had the same returns to endowments that White men have, we would expect the compensation of Black men to be $1,415 higher than what it is. Together, differences in endowments and returns to endowments account for about $2,500 of the total difference in compensation between Black men and White men, leaving $5,151 unexplained.

Figure B.7. Overall Decomposition of the Difference in Average Wages Between Black Men and White Men

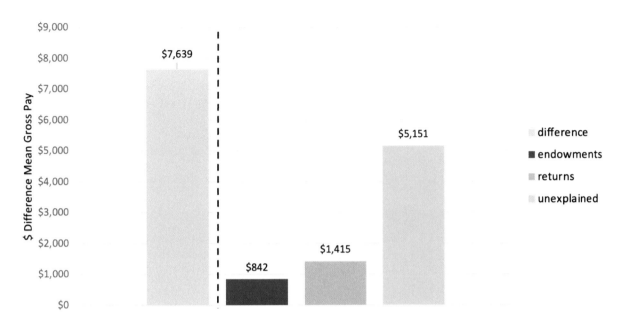

SOURCE: Authors' analysis of DMDC data on DoD civilian STEM workers (FYs 2018–2020) provided to RAND.

Figure B.8 provides the detailed breakdowns of the differences attributable to endowments and to returns. Although each overall difference is relatively small, the breakdown indicates that the small overall difference is the sum of some positive and some negative values. In terms of endowments (Panel B), Black men tend to be older, work for higher-paying agencies, and are in higher-paying locations than White men, on average. On the one hand, they have less tenure, lower education levels, and are in lower-paying pay plans. The difference in pay plans is particularly noticeable. If Black men were in the same pay plans as White men, we would expect their average compensation to increase by $1,774. Not only do Black men have lower tenure, they are also paid worse by tenure (see Panel B). If Black men were paid by tenure the way White men are, we would expect their average compensation to increase by $1,638. On the other hand, Black men are paid better by STEM category than are White men. Those are interesting findings, yet the unexplained portion of the pay difference dominates the explained piece. There are some unobserved differences between Black men and White men that result in a large difference in average pay.

Figure B.8. Detailed Decomposition of Average Wages Between Black Men and White Men

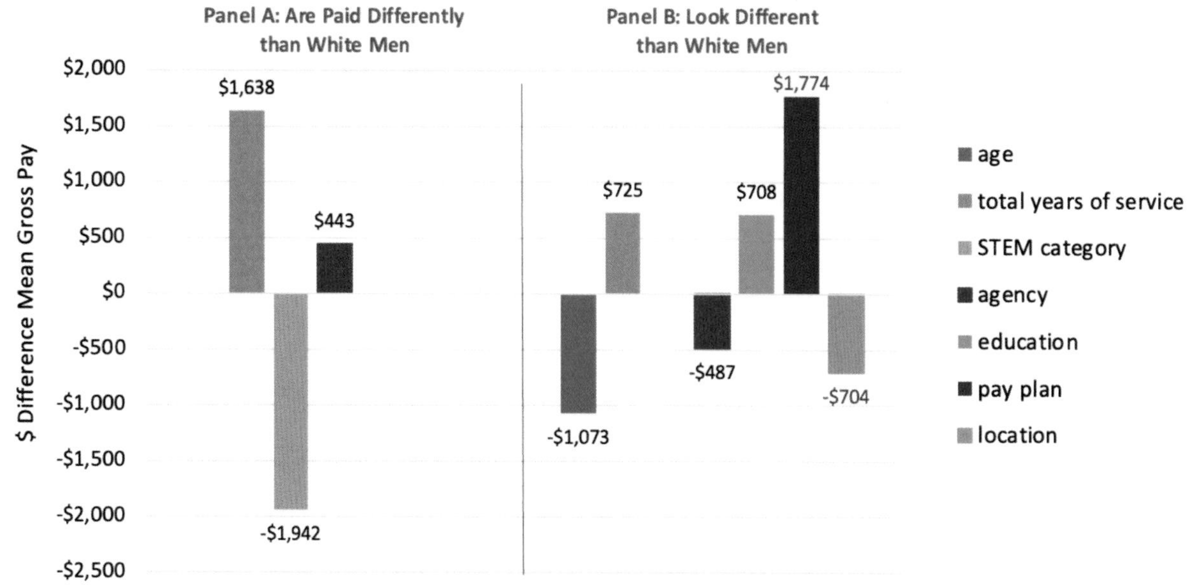

Panel A: Are Paid Differently than White Men

Panel B: Look Different than White Men

Legend:
- age
- total years of service
- STEM category
- agency
- education
- pay plan
- location

SOURCE: Authors' analysis of DMDC data on DoD civilian STEM workers (FYs 2018–2020) provided to RAND.

Black Women

The differences in compensation between Black women and White men break with almost all the previous trends. Black women have a raw difference in average compensation of just under $9,500 less than White men ($9,456 from regression, $9,330 from decomposition). Accounting for differences in observed characteristics slightly increases the pay gap. From the regression analysis, Black women have the lowest average wages compared with White men once differences in average compensation are considered. The decomposition results in Figure B.9 add to this story. Not only do differences in observed characteristics fail to explain the pay gap, but Black women also receive lower returns to endowments than White men. This difference in returns is the highest contributing factor to the overall compensation disparity. If Black women were paid like White men for their endowments, their average compensation would be $7,537 higher. The remaining unexplained portion of the pay gap is less than half that at $3,524. This means that—unlike for any other demographic group—we can explain most of the observed pay gap between White men and Black women through differences in returns. Although we cannot explain *why* such differences in returns exist, this is an important insight into the source of the pay gap for Black women compared with their White male counterparts.

Looking into the further details in Figure B.10, the difference in returns (Panel A) is driven by age and tenure. Black women are generally paid worse than White men of the same age and are generally paid worse than White men of the same tenure. In terms of endowments (Panel B), Black women tend to look better than White men in terms of age, tenure, education level, location, and agency. This is offset by differences in STEM category and pay plan: Black women fall into lower-paying STEM categories and worse-compensated pay plans.

Figure B.9. Overall Decomposition of the Difference in Average Wages Between Black Women and White Men

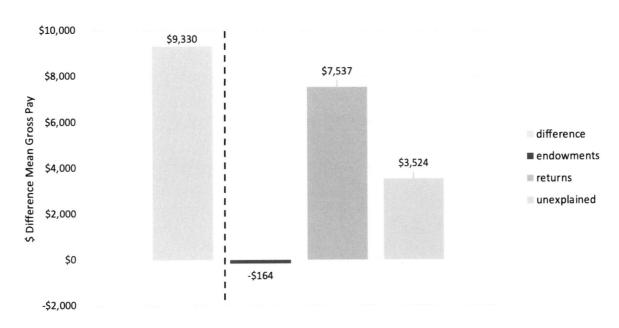

SOURCE: Authors' analysis of DMDC data on DoD civilian STEM workers (FYs 2018–2020) provided to RAND.

Figure B.10. Detailed Decomposition of Average Wages Between Black Women and White Men

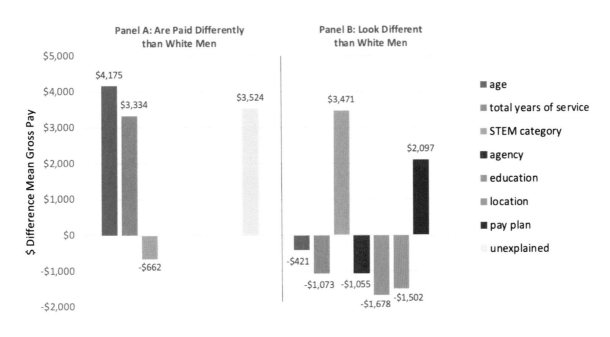

SOURCE: Authors' analysis of DMDC data on DoD civilian STEM workers (FYs 2018–2020) provided to RAND.

Asian Men

Asian men enjoy the highest unadjusted compensation of all the demographic groups considered; their average wages are more than $4,000 higher than those of White men. Figure B.11 presents the overall decomposition of the difference in average compensation of Asian men compared with White men. The driving force of the higher wages enjoyed by Asian men is their difference in endowments. If Asian men had similar endowments to White men, we would expect the average compensation of Asian men to fall by $7,210. Overall, the returns to endowments for Asian men and White men are similar. Unobserved factors unrelated to either observed endowments or the returns to those endowments remain and favor White men in comparison with Asian men, increasing their wages $2,347 on average. Although this unobserved difference is both substantial and statistically significant, it is the lowest level of unobserved difference of all the demographic groups considered.

Figure B.11. Overall Decomposition of the Difference in Average Wages Between Asian Men and White Men

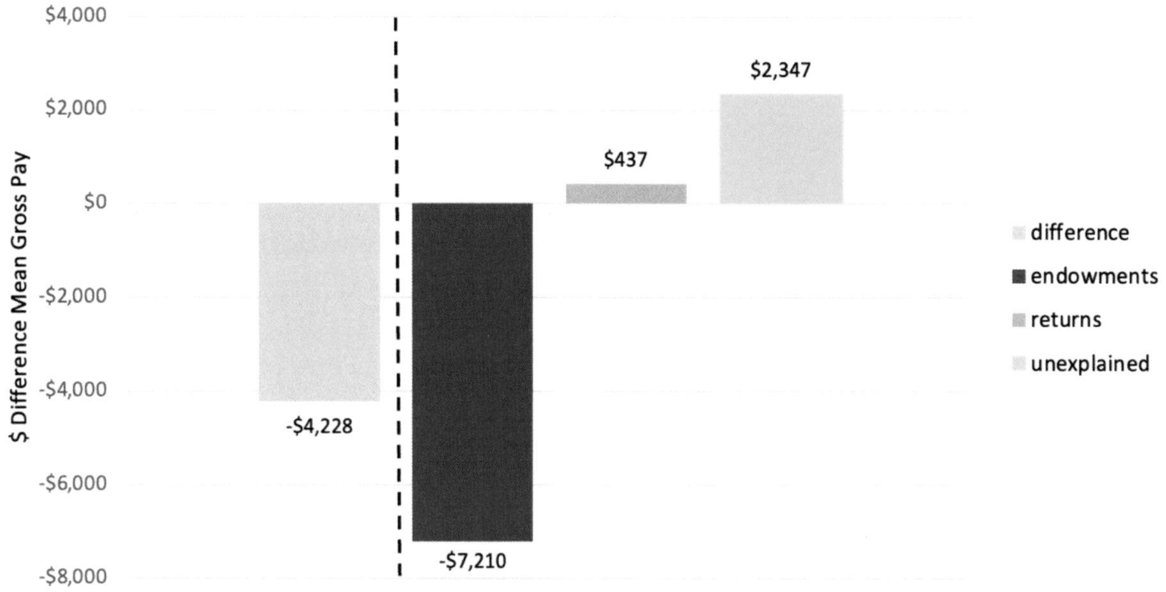

SOURCE: Authors' analysis of DMDC data on DoD civilian STEM workers (FYs 2018–2020) provided to RAND.

Figure B.12 provides the details of the endowment difference and the difference in returns. Although each difference in endowments contributes significantly to the overall compensation difference, education levels and location have the largest effects by far (Panel B). Differing education levels lead Asian men to earn averages wages that are $2,214 higher than White men, and Asian men tend to work in higher-paying locations. If Asian men were located where White men are located, we would expect Asian men to make almost $4,000 less, on average. Although Figure B.11 indicates that differences in returns to endowments have little overall effect on the wage gap between Asian men and White men, Panel A of Figure B.12 reveals that this overall effect is the combination of one large difference in returns in favor of White men (returns to age) and three moderately sized differences in returns in favor of Asian men (returns to tenure, STEM category, and pay plan). On the one hand, Asian men are paid worse by age than are White men, to the point that if Asian men were paid by age as White men are, Asian men would have compensation levels that are $3,830 higher on average. On the other hand, Asian men are paid better by tenure, STEM category, and pay plan, which results in $1,305, $1,535, and $2,012 higher average compensation levels than White men, respectively.

Figure B.12. Detailed Decomposition of Average Wages Between Asian Men and White Men

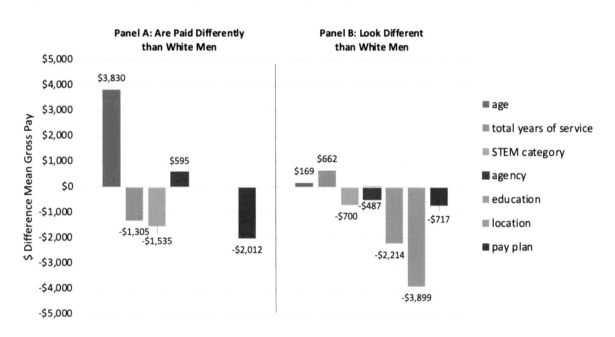

SOURCE: Authors' analysis of DMDC data on DoD civilian STEM workers (FYs 2018–2020) provided to RAND.

Asian Women

Asian women have the second-highest average compensation levels of the demographic groups considered, trailing only Asian men and leading White men by a raw difference of $769 (Figure B.13), and an annually adjusted difference of $625 (Figure 2.2 and Table 2.2 column [1]). According to the decomposition results of Figure B.13, Asian women have both higher-valued endowments and enjoy higher returns to their endowments than White men. Counterbalancing those factors is an unexplained compensation gap of $6,681 in favor of White men, which is the second-highest unexplained pay gap between White men and the other demographic groups considered here (Hispanic women have the largest). Aside from the worker and job characteristics considered in this analysis, there are factors differentiating Asian women from White men that are associated with White men earning over $6,500 more than Asian women, on average. Put another way, if the only factors that affected wages were the characteristics considered in the model, then using their endowments and returns, Asian women would make over $6,500 more than White men. Instead, Asian women earn about $770 more than White men.

Figure B.14 presents the details of the endowment piece and the returns piece of the overall decomposition, both of which are dwarfed by the unexplained part. Asian women are more highly educated than White men and live in higher-paying locations (Panel B). Asian women are paid worse by age, but better by tenure and STEM category (Panel A), resulting in better overall returns than White men (Figure B.13).

Figure B.13. Overall Decomposition of the Difference in Average Wages Between Asian Women and White Men

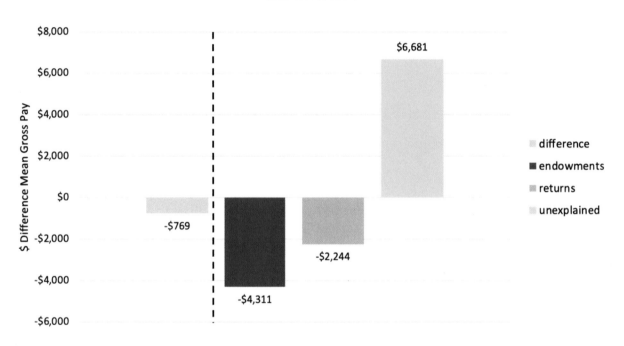

SOURCE: Authors' analysis of DMDC data on DoD civilian STEM workers (FYs 2018–2020) provided to RAND.

Figure B.14. Detailed Decomposition of Average Wages Between Asian Women and White Men

SOURCE: Authors' analysis of DMDC data on DoD civilian STEM workers (FYs 2018–2020) provided to RAND.

Appendix C. Supplemental Results for Play Plans and Occupations

This appendix provides additional information and supplemental results for the pay plan and pay grade analyses of Chapter 3 and the exploratory investigations within occupations of Chapter 4.

Additional Pay Plan Results

To analyze the differential use of pay plans by demographic group, we created a binary variable that takes value 0 if the worker is on a GG or GS pay plan and value 1 if the worker is on one of the alternative pay plans considered (DB, DP, DR, or NH). This is an indicator for being on an alternative plan to GG or GS. We then regress that alternative pay plan indicator on demographic-group indicators, as well as age, tenure, education category, location (U.S. state), STEM category, agency of employment, and the interaction between STEM category and agency of employment indicators.[36] We used logistic regression with standard errors clustered at the worker level and present average marginal effects. The average marginal effect is the estimated change in probability for either a one-unit increase in a continuous variable or for the level of a categorical variable compared with the reference group. For example, the average marginal effect for White women is the difference in predicted probability of being on an alternative pay plan for White women compared with the predicted probability for White men. The average marginal effect for age is the predicted change in probability for an additional year of age. In a logistic regression, the predicted probability is a function of all the variables. The average is taken over all the values of the other variables (Wooldridge, 2010). The average marginal effects are presented in Table C.1. The table omits location for space considerations, but it was included in the model. The average marginal effects for demographic groups are identical to the difference in probability presented in column 3 of Table 3.2.

---

[36] Including the interaction between STEM category and agency of employment instead of including each separately allows different agencies to have different patterns of pay plan use. For example, it allows the likelihood of being on an alternative pay plan for engineering workers to differ between the Navy and the Army.

Table C.1. Average Marginal Effects for Probability of Being on an Alternative Pay Plan

Model Indicators	Average Marginal Effect
Demographic group	
White women	−0.010***
Hispanic men	−0.012**
Hispanic women	−0.028***
Black men	−0.058***
Black women	−0.071***
Asian men	0.000
Asian women	0.003
STEM category	
Life science	−0.082***
IT and CS	0.024***
Engineering	0.079***
Social science	−
Agency	
Army	−0.106***
DoD	−
Navy	−0.120***
Education	
Master's	−0.192***
Bachelor's	−0.263***
Associate's	−0.450***
Technical college	−0.475***
No degree/some college	−0.447***
Age	−0.001***
Tenure	<0.001***

SOURCE: Authors' analysis of DMDC data on DoD civilian STEM workers (FYs 2016–2020) provided to RAND.
NOTES: ** $p < 0.01$ *** $p < 0.001$. Outcome is a binary indicator for being on an alternative pay plan to GG and GS. Logistic regression was used and results are marginal predictive probabilities (Adjusted Probability) and average marginal effects (last column). Note the average marginal effect is not estimated for DoD or for social science. This is because all workers in social science for DoD are on GG and GS pay plans, so the interaction term is subsumed in the constant term in the model.

The final part of analysis in Chapter 3 considered different pay grades within each pay plan. In this subsection, we considered GG and GS separately. We limited the pay grades considered and used number and percentage of workforce within that pay plan. For the alternative pay plans, we dropped pay grades with less than 1 percent of the DoD STEM workforce. For the GS pay plan, we dropped pay grades with fewer than 1,000 DoD civilian STEM workers. The full pay grade distribution within each pay plan is given in Table C.2.

Table C.2. Composition of Department of Defense STEM Workforce by Pay Grade for Each Pay Plan

Pay Plan	Pay Grade	N	Percentage	Cumulative Percentage
NH	NH-00	1	0.00	0.00
	NH-01	1	0.00	0.00
	NH-02	1,898	3.79	3.79
	NH-03	38,820	77.48	81.27
	NH-04	9,384	18.73	100.00
	Total	50,104	100.00	
GS	GS-02	6	0.00	0.00
	GS-03	90	0.02	0.02
	GS-04	403	0.09	0.11
	GS-05	2,611	0.60	0.72
	GS-06	1,041	0.24	0.96
	GS-07	14,048	3.23	4.19
	GS-08	1,682	0.39	4.58
	GS-09	40,597	9.35	13.93
	GS-10	5,363	1.24	15.16
	GS-11	88,832	20.46	35.62
	GS-12	147,343	33.93	69.55
	GS-13	96,667	22.26	91.81
	GS-14	27,410	6.31	98.12
	GS-15	8,158	1.88	100.00
	Total	434,251	100.00	
GG	GG-05	1	0.01	0.01
	GG-06	3	0.02	0.02
	GG-07	121	0.70	0.73
	GG-08	1	0.01	0.73
	GG-09	441	2.57	3.30
	GG-10	210	1.22	4.52
	GG-11	1,139	6.63	11.15
	GG-12	3,925	22.85	34.00
	GG-13	7,684	44.73	78.73
	GG-14	2,906	16.92	95.65
	GG-15	748	4.35	100.00
	Total	17,179	100.00	
DR	DR-01	692	5.52	5.52
	DR-02	5,204	41.49	47.01
	DR-03	4,353	34.71	81.72
	DR-04	2,292	18.27	99.99
	DR-05	1	0.01	100.00
	Total	12,542	100.00	
DP	DP-01	1	0.00	0.00
	DP-02	3,161	11.66	11.66
	DP-03	4,745	17.50	29.16
	DP-04	14,846	54.74	83.90
	DP-05	4,304	15.87	99.77

Pay Plan	Pay Grade	N	Percentage	Cumulative Percentage
	DP-06	62	0.23	100.00
	Total	27,119	100.00	
DB	DB-02	3,039	9.45	9.45
	DB-03	17,523	54.51	63.97
	DB-04	10,729	33.38	97.34
	DB-05	791	2.46	99.80
	DB-06	63	0.20	100.00
	Total	32,145	100.00	

SOURCE: Authors' analysis of DMDC data on DoD civilian STEM workers (FYs 2016–2020) provided to RAND.

Additional Occupational Explorations

Chapter 4 considered each demographic group separately and the limited analysis to the location, agency, STEM category with the most workers of that group. The average compensation for that group in that combination of location, agency, and STEM category was then compared with the average compensation of White men in the same location, agency, and STEM category.

Table C.3. Average Compensation by Demographic Group for GS-12 Workers by Location, Agency, and STEM Category Split by Pay Plan (Supplement to Table 4.3)

Group	Pay Plan	Group (%)	Compensation ($)	Compensation for White Men ($)	White Men (%)
White women	GG/GS	92.0	83,712	93,237	92.2
	DP	1.4	104,019	125,682	1.9
	NH	6.5	126,067	138,992	5.9
Hispanic men	GG/GS	69.9	88,521	98,200	61.6
	DP	29.7	89,766	105,408	37.6
	NH	0.5	131,272	131,573	0.8
Hispanic women	GG/GS	63.1	84,828	98,200	61.6
	DP	35.4	88,939	105,408	37.6
	NH	1.5	116,969	131,573	0.8
Black men	GG/GS	95.9	85,637	93,237	92.2
	DP	1.0	76,062	125,682	1.9
	NH	3.1	118,591	138,992	5.9
Black women	GG/GS	91.3	85,202	103,327	86.9
	DP	0.7	117,091	131,975	1.6
	NH	8.0	117,160	124,105	11.5
Asian men	GG/GS	65.0	96,139	98,200	61.6
	DP	32.0	106,021	105,408	37.6
	NH	3.0	118,318	131,573	0.8
Asian women	GG/GS	99.9	99,455	97,675	99.8
	NH	0.1	116,899	114,561	0.2

SOURCE: Authors' analysis of DMDC data on DoD civilian STEM workers (FYs 2016–2020) provided to RAND.

Abbreviations

AcqDemo	Acquisition Workforce Demonstration Project
CPS	Current Population Survey
CS	computer science
DB	demonstration engineer and scientist
demo	demonstration project
DEIA	diversity, equity, inclusion, and accessibility
DMDC	Defense Manpower Data Center
DoD	U.S. Department of Defense
DP	demonstration professional
DR	demonstration Air Force scientist and engineer
FY	fiscal year
GG	General Government
GS	General Schedule
IT	information technology
Lab Demo	Science and Technology Reinvention Laboratory Demonstration Project
ND	demonstration, scientific and engineering professional
NH	business management and technical management professional
NSRD	RAND National Security Research Division
OPM	U.S. Office of Personnel Management
OSD	U.S. Office of the Secretary of Defense
STEM	science, technology, engineering, and mathematics

References

Asch, Beth J., Michael G. Mattock, and James Hosek, *How Do Federal Civilian Pay Freezes and Retirement Plan Changes Affect Employee Retention in the Department of Defense?* RAND Corporation, RR-678-OSD, 2014. As of September 30, 2022:
https://www.rand.org/pubs/research_reports/RR678.html

Blau, Francine D., and Lawrence M. Kahn, "The Gender Wage Gap: Extent, Trends, and Explanations," *Journal of Economic Literature*, Vol. 55, No. 3, September 2017.

Biden, Joseph R., Jr., "Executive Order on Diversity, Equity, Inclusion, and Accessibility in the Federal Workforce," Executive Office of the President, June 25, 2021.

DoD—*See* U.S. Department of Defense.

Edwards, Kathryn A., Maria McCollester, Brian Phillips, Hannah Acheson-Field, Isabel Leamon, Noah Johnson, and Maria C. Lytell, *Compensation and Benefits for Science, Technology, Engineering, and Mathematics (STEM) Workers: A Comparison of the Federal Government and the Private Sector*, RAND Corporation, RR-4267-OSD, 2021. As of June 1, 2022:
https://www.rand.org/pubs/research_reports/RR4267.html

FEDweek, "DoD Plans Greater Pay Flexibilities in Demonstration Project Labs," *Federal Manager's Daily Report*, May 12, 2022. As of October 3, 2022:
https://www.fedweek.com/federal-managers-daily-report/dod-plans-greater-pay-flexibilities-in-demonstration-project-labs

Groeber, Ginger, Paul W. Mayberry, Brandon Crosby, Mark Doboga, Samantha E. DiNicola, Caitlin Lee, and Ellen E. Tunstall, *Federal Civilian Workforce Hiring, Recruitment, and Related Compensation Practices for the Twenty-First Century: Review of Federal HR Demonstration Projects and Alternative Personnel Systems to Identify Best Practices and Lessons Learned*, RAND Corporation, RR-3168-OSD, 2020. As of October 3, 2022:
https://www.rand.org/pubs/research_reports/RR3168.html

Jann, Ben, "The Blinder–Oaxaca Decomposition for Linear Regression Models," *Stata Journal*, Vol. 8, No. 4, 2008.

Keller, Kirsten M., Maria C. Lytell, David Schulker, Kimberly Curry Hall, Louis T. Mariano, John S. Crown, Miriam Matthews, Brandon Crosby, Lisa Saum-Manning, Douglas Yeung, Leslie Adrienne Payne, Felix Knutson, and Leann Caudill, *Advancement and Retention Barriers in the U.S. Air Force Civilian White Collar Workforce: Implications for Demographic Diversity*, RAND Corporation, RR-2643-AF, 2020. As of September 26, 2022: https://www.rand.org/pubs/research_reports/RR2643.html

Kim, ChangHwan, "Decomposing the Change in the Wage Gap Between White and Black Men over Time, 1980–2005: An Extension of the Blinder-Oaxaca Decomposition Method," *Sociological Methods and Research*, Vol. 38, No. 4, 2010.

Lewis, Jennifer Lamping, Laura Werber, Cameron Wright, Irina Elena Danescu, Jessica Hwang, and Lindsay Daugherty, *2016 Assessment of the Civilian Acquisition Workforce Personnel Demonstration Project*, RAND Corporation, RR-1783-OSD, 2017. As of September 30, 2022: https://www.rand.org/pubs/research_reports/RR1783.html

Matthews, Miriam, Bruce R. Orvis, David Schulker, Kimberly Curry Hall, Abigail Haddad, Stefan Zavislan, and Nelson Lim, *Hispanic Representation in the Department of Defense Civilian Workforce: Trend and Barrier Analysis*, RAND Corporation, RR-1699-OSD, 2017. As of September 30, 2022: https://www.rand.org/pubs/research_reports/RR1699.html

Naylor, Brian, "Trump Lifting Federal Hiring Freeze," NPR, April 12, 2017.

Okrent, Abigail, and Amy Burke, *The STEM Labor Force of Today: Scientists, Engineers, and Skilled Technical Workers*, National Science Board, NSB-2021-2, August 31, 2021. As of January 20, 2023: https://ncses.nsf.gov/pubs/nsb20212

OPM—*See* U.S. Office of Personnel Management.

U.S. Census Bureau, "Methodology," webpage, updated November 19, 2021. As of October 3, 2022: https://www.census.gov/programs-surveys/cps/technical-documentation/methodology.html

U.S. Census Bureau, "Current Population Survey (CPS)," webpage, updated December 13, 2022. As of October 3, 2022: https://www.census.gov/programs-surveys/cps.html

U.S. Department of Defense, "Acquisition Career Management in the 4th Estate," June 20, 2019. As of October 3, 2022: https://www.defense.gov/News/Feature-Stories/story/Article/1882166/acquisition-career-management-in-the-4th-estate/

U.S. Department of Defense, "Department of Defense Science and Technology Reinvention Laboratory (STRL) Personnel Demonstration (Demo) Project in the U.S. Army Research Institute for the Behavioral and Social Sciences (ARI)," *Federal Register*, Vol. 85, No. 229, November 27, 2020.

U.S. Office of Personnel Management, "Fact Sheet: Pay Plans," webpage, undated-b. As of October 3, 2022:
https://www.opm.gov/policy-data-oversight/pay-leave/pay-administration/fact-sheets/pay-plans/

U.S. Office of Personnel Management, "Federal Employee Viewpoint Survey: About," webpage, undated-a. As of November 11, 2022:
https://www.opm.gov/fevs/about/

U.S. Office of Personnel Management, *Handbook of Occupational Groups and Families*, December 2018.

Wooldridge, Jeffrey M., *Econometric Analysis of Cross Section and Panel Data*, MIT Press, 2010.